Chemical Principles in the Laboratory

FOURTH EDITION

Chemical Principles in the Laboratory

FOURTH EDITION

Emil J. Slowinski

Professor of Chemistry
Macalester College
St. Paul, Minnesota

Wayne C. Wolsey

Professor of Chemistry
Macalester College
St. Paul, Minnesota

William L. Masterton

Professor of Chemistry
University of Connecticut
Storrs, Connecticut

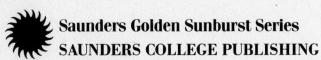

Saunders Golden Sunburst Series

SAUNDERS COLLEGE PUBLISHING

Philadelphia   New York   Chicago
San Francisco   Montreal   Toronto
London   Sydney   Tokyo   Mexico City
Rio de Janeiro   Madrid

Address orders to:
383 Madison Avenue
New York, NY 10017

Address editorial correspondence to:
West Washington Square
Philadelphia, PA 19105

Text Typeface: 10/12 Caledonia
Compositor: York Graphic Services, Inc.
Acquisitions Editor: John Vondeling
Project Editor: Joanne Fraser
Copyeditor: Mark Hobbs
Art Director: Carol C. Bleistine
Art/Design Assistant: Virginia A. Bollard
Text Design: Nancy E. J. Grossman
Cover Design: Lawrence R. Didona
Text Artwork: Larry Ward
Production Manager: Tim Frelick
Assistant Production Manager: Maureen Iannuzzi

Cover credit: Pouring Steel. © Alvin Upitis/THE IMAGE BANK

CHEMICAL PRINCIPLES IN THE LABORATORY   ISBN 0-03-070754-4

3456   019   987654321

CBS COLLEGE PUBLISHING
Saunders College Publishing
Holt, Rinehart and Winston
The Dryden Press

PREFACE

In spite of its many successful theories, chemistry remains, and probably always will remain, an experimental science. Most of the research, both in universities and in industry, is done in the laboratory rather than in the office or computing room, and it behooves the young student of chemistry to devote a substantial portion of time to the experimental aspects of the subject. It is not easy to become a good experimentalist, and it must be admitted that some chemists are not always effective in the laboratory. Yet even those chemists who do not work full-time in the laboratory must be familiar with the available experimental methods, the proper design of experiments, and the interpretation of experimental results. As beginning chemists, students will find that their efforts in the laboratory will be rewarded by a better understanding of the concepts of chemistry as well as an appreciation of what is required in the way of technique and interpretation if one is to be able to find or demonstrate any chemically significant relations.

In writing this manual, the authors have attempted to illustrate many of the established principles of chemistry with experiments that are as interesting and challenging as possible. For the most part, the experimental procedures and methods for processing data are described in great detail, so that students of widely varying backgrounds and abilities will be able to see how to perform the experiments properly and how to interpret them. Many of the experiments in this manual have not appeared elsewhere, and were developed and tested in the general chemistry laboratories of Macalester College and the University of Connecticut. We have included more experiments than can be conveniently done in a year in the usual laboratory program, so instructors may select experiments in a flexible way to meet the needs of their particular courses. In many of the experiments, unknowns are assigned to students to ensure their working independently and to add a measure of realism to the laboratory experience.

In preparation for writing this fourth edition, we sent out questionnaires to current users of the manual, asking for comments about the experiments and suggestions for improvements or changes. These comments have been carefully considered, and many of the suggestions have been incorporated in the manual. Several of the experiments have improved procedures, some experiments from the alternate edition of the manual have been included, a few experiments have been dropped, and some experiments from earlier editions have been reinstated in this one. The order of the experiments makes them compatible with the sixth edition of our general chemistry text, *Chemical Principles*. We believe, however, that the overall set of experiments should be appropriate for use with most of the modern texts in general chemistry.

As in earlier editions, we include with each experiment an advance study assignment, designed to help the student prepare for the laboratory session. The questions in the advance study assignments include in nearly every case data similar to those the student will obtain in the experiment. Directions for treating the data have been given in great detail, so if the student properly completes the advance study assignments, he or she should have no trouble in performing the experiments or in processing the data.

We have selected the experiments with some regard to cost, since both chemicals and equipment are expensive. In the teacher's guide, we note the cost of the chemicals per student for each experiment. We have attempted to keep the experiments safe, have dropped toxic reagents or suspected carcinogens where that was feasible, and have included safety warnings in experimental procedures where they seemed indicated.

The system of units we have used in this manual is essentially the same as in the third edition.

For the most part, the system is SI, but we have kept mm Hg and atmospheres for pressure measurements, and usually use milliliters and liters for expressing volumes.

The authors are grateful to those who replied to our questionnaire and for the support of those who have used our earlier editions. We would much appreciate any comments you may have about this version of the manual.

E. J. SLOWINSKI

W. C. WOLSEY

W. L. MASTERTON

SAFETY IN THE LABORATORY

Read this section before performing any of the experiments in this manual.

A chemistry laboratory can be, and should be, a safe place in which to work. Yet each year in academic and industrial laboratories accidents occur that in some cases injure seriously, or kill, chemists. Most of these accidents could have been foreseen and prevented, had the chemists involved used the proper judgment and taken proper precautions.

The experiments you will be performing have been selected at least in part because they can be done safely. Instructions in the procedures should be followed carefully and in the order given. Sometimes even a change in concentration of one reagent is sufficient to change the conditions of a chemical reaction so as to make it occur in a different way, perhaps at a highly accelerated rate. So, do not deviate from the procedure given in the manual when performing experiments unless specifically told to do so by your instructor.

Eye Protection. One of the simplest, and most important, things you can do in the laboratory to avoid injury is to protect your eyes by routinely wearing safety glasses. Your instructor will tell you what eye protection to use, and you should use it. Goggles worn up on the hair may be attractive, but they are not protective. It is not advisable to wear contact lenses in the laboratory.

Chemical Reagents. Chemicals in general are toxic materials. This means that they can act as poisons or carcinogens (causes of cancer) if they get into your digestive or respiratory system. Never taste a chemical substance, and avoid getting any chemical on your skin. If that should happen, wash it off promptly with plenty of water. Also, wash your face and hands when you are through working in the laboratory. Do not pipet by mouth; when pipetting, use a rubber bulb or other device to suck up the liquid. Avoid breathing vapors given off by reagents or reactions. If directed to smell a vapor, do so cautiously. Use the hood when the directions call for it.

Some reagents, such as concentrated acids or bases, or bromine, are caustic, which means that they can cause chemical burns on your skin and eat through your clothing. Where such reagents are being used, we note the potential danger with a *CAUTION* sign at that point in the procedure. Be particularly careful when carrying out that step. Always read the label on a reagent bottle before using it; there is a lot of difference between the properties of 1 M H_2SO_4 and those of concentrated (18 M) H_2SO_4.

A few of the chemicals we use are flammable. These include hexane, ethanol, and acetone. Keep your Bunsen burner well away from any open beakers containing such chemicals, and be careful not to spill them on the laboratory bench, where they might easily get ignited.

When disposing of the chemical products from an experiment, use good judgment. Some dilute, nontoxic solutions can be poured down the sink and flushed with water. Insoluble or toxic materials should be put in the waste crocks provided for that purpose. If you are in doubt about how to dispose of a chemical mixture, ask your instructor.

Safety Equipment.   In the laboratory there are various pieces of safety equipment, which may include a safety shower, an eye wash fountain, a fire extinguisher, and a fire blanket. Learn where these items are, so that you will not have to look all over if you ever need them in a hurry.

Laboratory Attire.   Come to the laboratory in sensible clothing. Long, flowing robes are out, as are bare feet. Sandals and open-toed shoes offer less protection than regular shoes. Keep long hair tied back, out of the way of flames and reagents.

If an Accident Occurs.   During the laboratory course a few accidents will probably occur. For the most part these will not be serious, and might involve a spilled reagent, a beaker of hot water that gets tipped over, a dropped test tube, or a small fire.

A common response in such a situation is panic. A student may respond to an otherwise minor accident by doing something irrational, like running from the laboratory when the remedy for the accident is close at hand. If an accident happens to another student, watch for signs of panic and tell the student what to do; if it seems necessary, help him or her do it. Call the instructor for assistance.

Chemical spills are best handled by washing the area quickly with water from the nearest sink. Use the eye wash fountain if you get something in your eye. In case of a severe chemical spill on your clothing or shoes, use the emergency shower and take off the affected clothing. In case of a fire in a beaker, on a bench, or on your clothing or that of another student, do not panic and run. Smother the fire with an extinguisher, with a blanket, or with water, as seems most appropriate at the time. If the fire is in a piece of equipment or on the lab bench and does not appear to require instant action, have your instructor put the fire out. If you cut yourself on a piece of broken glass, tell your instructor, who will assist you in treating it.

A Message to Foreign Students.   Many students from foreign countries take courses in chemistry before they are completely fluent in English. If you are such a student, it may be that in some experiments you will be given directions that you do not completely understand. If that happens, do not try to do that part of the experiment by simply doing what the student next to you seems to be doing. Ask that student, or the instructor, what the confusing word or phrase means, and when you understand what you should do, go ahead. You will soon learn the language well enough, but until you feel comfortable with it, do not hesitate to ask others to help you with unfamiliar phrases and expressions.

Although we have spent considerable time here describing some of the things you should be concerned with in the laboratory from a safety point of view, this does not mean you should work in the laboratory in fear and trepidation. Chemistry is not a dangerous activity when practiced properly. Chemists as a group live longer than other professionals, in spite of their exposure to potentially dangerous chemicals. In this manual we have attempted to describe safe procedures and to employ chemicals that are safe when used properly. Many thousands of students have performed the experiments without having accidents, so you can too. However, we authors cannot be in the laboratory when you carry out the experiments to be sure that you observe the necessary precautions. You and your laboratory supervisor must, therefore, see to it that the experiments are done properly and assume responsibility for any accidents or injuries that may occur.

CONTENTS

EXPERIMENT

1 • The Densities of Liquids and Solids

One of the fundamental properties of any sample of matter is its density, which is its mass per unit of volume. The density of water is exactly 1.00000 g/cm³ at 4°C and is slightly less than one at room temperature (0.9970 g/cm³ at 25°C). Densities of liquids and solids range from values less than that of water to values considerably greater than that of water. Osmium metal has a density of 22.5 g/cm³ and is probably the densest material known at ordinary pressures.

In any density determination, two quantities must be determined—the mass and the volume of a given quantity of matter. The mass can easily be determined by weighing a sample of the substance on a balance. The quantity we usually think of as "weight" is really the mass of a substance. In the process of "weighing" we find the mass, taken from a standard set of masses, that experiences the same gravitational force as that experienced by the given quantity of matter we are weighing. The mass of a sample of liquid in a container can be found by taking the difference between the mass of the container plus the liquid and the mass of the empty container.

The volume of a liquid can easily be determined by means of a calibrated container. In the laboratory a graduated cylinder is often used for routine measurements of volume. Accurate measurement of liquid volume is made by using a pycnometer, which is simply a container having a precisely definable volume. The volume of a solid can be determined by direct measurement if the solid has a regular geometrical shape. Such is not usually the case, however, with ordinary solid samples. A convenient way to determine the volume of a solid is to measure accurately the volume of liquid displaced when an amount of the solid is immersed in the liquid. The volume of the solid will equal the volume of liquid which it displaces.

In this experiment we will determine the density of a liquid and a solid by the procedure we have outlined. First we weigh an empty flask and its stopper. We then fill the flask completely with water, measuring the mass of the filled stoppered flask. From the difference in these two masses we find the mass of water and then, from the known density of water, we determine the volume of the flask. We empty and dry the flask, fill it with an unknown liquid, and weigh again. From the mass of the liquid and the volume of the flask we find the density of the liquid. To determine the density of an unknown solid metal, we add the metal to the dry empty flask and weigh. This allows us to find the mass of the metal. We then fill the flask with water, leaving the metal in the flask, and weigh again. The increase in mass is that of the added water; from that increase, and the density of water, we calculate the volume of water we added. The volume of the metal must equal the volume of the flask minus the volume of water. From the mass and volume of the metal we calculate its density. The calculations involved are outlined in detail in the Advance Study Assignment.

EXPERIMENTAL PROCEDURE

A. Mass of a Slug. After you are shown how to operate the analytical balance in your laboratory, obtain a numbered metal slug from your instructor. Weigh it on the balance to the nearest 0.001 g. Record the mass and the number of the slug and report it to your instructor. When he has approved your weighing, go to the stockroom and obtain a glass-stoppered flask, which will serve as a pycnometer, and samples of an unknown liquid and an unknown metal.

B. Density of a Liquid. If your flask is not clean and dry, clean it with soap and water, rinse it with a few cubic centimeters of acetone, and dry it by letting it stand for a few minutes in the air or by *gently* blowing compressed air into it for a few moments.

Weigh the dry flask with its stopper on the analytical balance, or the toploading balance if so directed, to the nearest milligram. Fill the flask with distilled water until the liquid level is nearly to the *top* of the ground surface in the neck. Put the stopper in the flask in order to drive out *all* the air and any excess water. Work the stopper gently into the flask, so that it is firmly seated in position. Wipe any water from the outside of the flask with a towel and soak up all excess water from around the top of the stopper.

Again weigh the flask, which should be completely dry on the outside and full of water, to the nearest milligram. Given the density of water at the temperature of the laboratory and the mass of water in the flask, you should be able to determine the volume of the flask very precisely. Empty the flask, dry it, and fill it with your unknown liquid. Stopper and dry the flask as you did when working with the water and then weigh the stoppered flask full of the unknown liquid, making sure its surface is dry. This measurement, used in conjunction with those you made previously, will allow you to find accurately the density of your unknown liquid.

C.   Density of a Solid.   Pour your sample of liquid from the flask into its container. Rinse the flask with a small amount of acetone and dry it thoroughly. Add small chunks of the metal sample to the flask until the flask is about half full. Weigh the flask, with its stopper and the metal, to the nearest milligram. You should have at least 50 g of metal in the flask.

Leaving the metal in the flask, fill the flask with water and then replace the stopper. Roll the metal around in the flask to make sure that no air remains between the metal pieces. Refill the flask if necessary, and then weigh the dry, stoppered flask full of water plus the metal sample. Properly done, the measurements you have made in this experiment will allow a calculation of the density of your metal sample that will be accurate to about 0.1 per cent.

Pour the water from the flask. Put the metal in its container. Dry the flask and return it with its stopper and your metal sample to the stockroom.

DATA AND CALCULATIONS: Densities of Liquids and Solids

$$D = \frac{m}{V}$$

.9975

Metal slug no. __14__   Mass of slug __5.028__ g

Unknown liquid no. __2__   Unknown solid no. _____

		Trial I	Trial II
Density of unknown liquid			
Mass of empty flask plus stopper	23.1°C	44.07 g	22.9° @ 44.28 g
Mass of stoppered flask plus water		99.18 g	99.49 g
Mass of stoppered flask plus liquid		87.99 g	87.81 g
Mass of water		_____ g	_____
Volume of flask (density of H_2O at 25°C, 0.9970 g/cm³; at 20°C, 0.9982 g/cm³)		_____ cm³	_____
Mass of liquid		_____ g	_____
Density of liquid		_____ g/cm³	_____
To how many significant figures can the liquid density be properly reported?		_____	_____
Density of unknown metal			
Mass of stoppered flask plus metal		_____ g	
Mass of stoppered flask plus metal plus water		_____ g	
Mass of metal		_____ g	
Mass of water		_____ g	
Volume of water		_____ cm³	
Volume of metal		_____ cm³	
Density of metal		_____ g/cm³	

Would you expect the per cent error in the metal density to be higher or lower than the per cent error in the liquid density as obtained in this experiment? _____

Why?

Weigh a beaker dry.
Use a ~~buret~~ buret to obtain the
desired volume of ~~liquid~~. Weigh the beaker
again when it ~~O~~ ~~the~~ is full of the liquid.

mass of beaker   49.45   g   49.45   g
mass of beaker + ~~wat~~ liquid 69.02   g   23.4°c 69.07
mass of ~~wa~~ liquid   _____ g   _____ g
volume of liquid   _____ cm³ _____ cm³
Density of liquid   _____ g/cm³ _____ g/cm³

EXPERIMENT

2 • Resolution of Matter into Pure Substances, I. Fractional Crystallization

One of the important problems faced by chemists is that of determining the nature and state of purity of the substances with which they work. In order to perform meaningful experiments, chemists must ordinarily use essentially pure substances, which are often prepared by separation from complex mixtures.

In principle the separation of a mixture into its component substances can be accomplished by carrying the mixture through one or more physical changes, experimental operations in which the nature of the components remains unchanged. Because the physical properties of various pure substances are different, physical changes frequently allow an enrichment of one or more substances in one of the fractions that is obtained during the change. Many physical changes can be used to accomplish the resolution of a mixture, but in this experiment we will restrict our attention to one of the simpler ones in common use, namely, fractional crystallization.

The solubilities of solid substances in different kinds of liquid solvents vary widely. Some substances are essentially insoluble in all known solvents; the materials we classify as macromolecular are typical examples. Most materials are noticeably soluble in one or more solvents. Those substances that we call salts often have very appreciable solubility in water but relatively little solubility in any other liquids. Organic compounds, whose molecules contain carbon and hydrogen atoms as their main constituents, are often soluble in organic liquids such as benzene or carbon tetrachloride.

We also often find that the solubility of a given substance in a liquid is sharply dependent on temperature. Most substances are more soluble in a given solvent at high temperatures than at low temperatures, although there are some materials whose solubility is practically temperature-independent and a few others that become less soluble as temperature increases.

By taking advantage of the differences in solubility of different substances we often find it possible to separate the components of a mixture in essentially pure form.

In this experiment you will be given a sample containing silicon carbide, potassium nitrate, and copper sulfate. Your problem will be to separate two pure components from the mixture, using water as the solvent. Silicon carbide, SiC, is a black, very hard material; it is the classic abrasive, and completely insoluble in water. Potassium nitrate, KNO_3, and copper sulfate, $CuSO_4 \cdot 5\ H_2O$, are water-soluble ionic substances, with different solubilities at different temperatures, as indicated in Figure 2.1. The copper sulfate we will use is blue in its crystalline hydrate and in solution. The solubility of the hydrate increases fairly rapidly with temperature. Potassium nitrate is a white solid, colorless in solution. Its solubility increases about twenty-fold between 0°C and 100°C.

Given a mixture containing roughly equal amounts of SiC and KNO_3 and a small amount of $CuSO_4 \cdot 5\ H_2O$, we separate out the silicon carbide first. This is done by simply stirring the mixture with water, which dissolves all of the potassium nitrate and copper sulfate in the mixture. The insoluble silicon carbide remains behind and is filtered off.

The solution obtained after filtration contains KNO_3 and $CuSO_4$ in a rather large amount of water. Some of the water is removed by boiling, and then the solution is cooled to 0°C. At that point the KNO_3 is not very soluble, and most of it crystallizes from solution. Since $CuSO_4$ is not present in large amount, its solubility is not exceeded and it remains in solution. The solid KNO_3 is separated from the solution by filtration. This procedure, by which a substance can be separated from an impurity, is called fractional crystallization.

The solid potassium nitrate one recovers is contaminated by a small amount of copper sulfate. By redissolving the solid in a minimum amount of hot water, and then cooling and recrystallizing the KNO_3, its purity can be very markedly increased. The purity can be established by the intensity of the color produced by the copper impurity when treated with ammonia, NH_3.

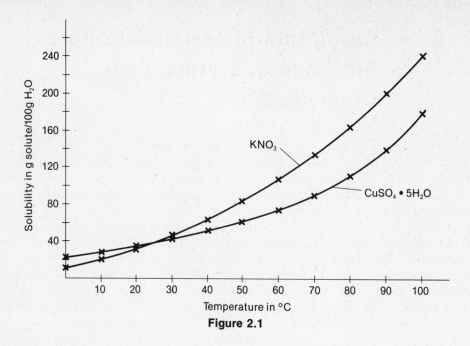

Figure 2.1

EXPERIMENTAL PROCEDURE

Obtain from the stockroom a Buchner funnel, a suction flask, and a sample (about 25 grams) of your unknown solid mixture.

Weigh a 150-ml beaker (to ± 0.1 g) on a top-loading balance. Add the sample and weigh again. Then add about 50 ml of distilled water, which will be enough to dissolve the soluble solids.

Separation of SiC. Support the beaker with its solution on a piece of wire gauze on an iron ring. Warm gently to about 40°C, while stirring the mixture. When the blue and white solids are all in solution, pour it into a Buchner funnel while gentle suction is being applied (see Fig. 2.2). Transfer as much as you can of the solid carbide to the funnel with your rubber policeman. Transfer the blue filtrate to the (cleaned) 150-ml beaker and add 15 drops of 6 M HNO_3, nitric acid, which will help ensure that the copper sulfate remains in solution in later steps. Reassemble the funnel, apply suction, and wash the SiC on the filter paper with distilled water. Continue the suction for a few minutes to dry the SiC. Turn off the suction, and, using your spatula, lift the filter

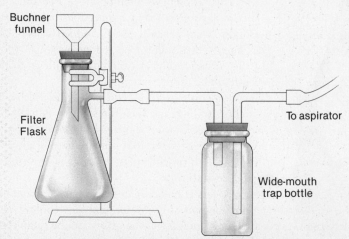

Figure 2.2. To operate the Buchner funnel, put a piece of circular filter paper in the funnel. Turn on suction and spray filter paper with distilled water from wash bottle. Keep suction on while filtering sample.

paper and the SiC crystals from the funnel, and put the paper on the lab bench so that the crystals may dry in the air. Prepare some ice-cold distilled water by putting your wash bottle in an ice-water bath.

Separation of KNO_3.   Heat the blue filtrate in the beaker to the boiling point, and then boil gently until white crystals of KNO_3 are visible in the liquid. CAUTION: The hot liquid will have a tendency to bump, so do not heat it too strongly. When white crystals are clearly apparent (the solution may appear cloudy at that point) stop heating and add 5 ml distilled water to the solution. Stir the mixture with a glass stirring rod to dissolve the solids, including any on the wall; if necessary, warm the solution but do not boil it.

Cool the solution to room temperature in a water bath, and then to about 0°C in an ice bath. White crystals of KNO_3 will come out of solution. If mixture becomes essentially solid, add an additional 5 ml distilled water. Stir the cold slurry of crystals for several minutes. Assemble the Buchner funnel; chill it by adding 100 ml ice-cold distilled water and, after about a minute, drawing the water through with suction. Filter the KNO_3 slurry through the cold Buchner funnel. Your rubber policeman will be helpful when transferring the last of the crystals. Press the crystals dry with a clean piece of filter paper, and continue to apply suction for about 30 seconds. Turn off the suction.

Wait a few moments, and then, without applying suction, add a *small* amount of ice-cold distilled water from your wash bottle to the funnel, just enough to wet the crystals. Let the cold liquid remain in contact with the crystals for about ten seconds, and then apply suction, for about a minute or so, to dry the purified KNO_3 crystals. Lift the filter paper and the crystals from the funnel and put the crystals on a dry piece of filter paper on the lab bench.

By this procedure you have separated most of the KNO_3 in your sample from the $CuSO_4$, which is present in the solution in the suction flask. This solution may now be discarded, so pour it into the waste crock, not down the sink.

Analysis of the Purity of the KNO_3.   The KNO_3 crystals you have prepared contain a small amount of $CuSO_4$ as an impurity. To find the amount of $CuSO_4$ present, weigh out 0.5 g (±0.1 g) of the crystals into a weighed 50-ml beaker. Dissolve the crystals in 3 ml distilled water, and then add 3 ml 6 M NH_3, ammonia. The copper impurity will form a blue solution in the NH_3. Compare the intensity of the blue color with that in a series of standard solutions prepared by your instructor. Estimate the relative concentration of $CuSO_4 \cdot 5 H_2O$ in your product.

Recrystallization of the KNO_3.   Clean and dry the 150-ml beaker you used in the first part of the experiment, and to it add your sample of KNO_3. Weigh to ±0.1 g on the top-loading balance. Given the mass of KNO_3, use Figure 2.1 to estimate the amount of water needed to dissolve the solid at 100°C. Add twice that amount of distilled water to the sample and heat to boiling. If necessary, add 1 ml of distilled water to complete the solution process.

Cool the solution to room temperature and then to 0°C in an ice bath as before. After stirring the crystals for several minutes, filter the slurry through an ice-cold Buchner funnel, transferring as much of the solid as possible to the funnel. Press the crystals dry with a piece of filter paper and continue to apply suction for a minute or so. Lift the filter paper from the funnel and put it on the lab bench. Transfer the solid to a piece of weighed dry filter paper to facilitate drying (your instructor will tell you the mass of the paper).

Determine the amount of $CuSO_4$ impurity in your recrystallized sample as you did with the first batch, and record that value. The recrystallization should have very significantly increased the purity of your KNO_3.

Weigh your dry SiC and the remaining KNO_3 crystals on their pieces of filter paper, to ±0.1 g. Show your samples of SiC and KNO_3 to your instructor for evaluation.

DATA AND CALCULATIONS: Resolution of Pure Substances, I. Fractional Crystallization

	Original sample	Recrystallized sample
Unknown no. _____		
Mass of 150-ml beaker	_____ g	
Mass of sample plus beaker	_____ g	
Mass of sample	_____ g	
Mass of 50-ml beaker	_____ g	
Mass of 50-ml beaker plus KNO_3	_____ g	_____ g
Mass of KNO_3 used in analysis	_____ g	_____ g
% $CuSO_4 \cdot 5\ H_2O$ present in KNO_3	_____ %	_____ %
Mass of SiC plus filter paper	_____ g	
Mass of filter paper (furnished value)	_____ g	
Mass of SiC in sample	_____ g	
Per cent SiC in sample	_____ %	
Mass of recrystallized KNO_3 plus paper		_____ g
Mass of recrystallized KNO_3		_____ g
% of sample recovered as pure KNO_3		_____ %

ADVANCE STUDY ASSIGNMENT: Resolution of Matter into Pure Substances, I.
Fractional Crystallization

1. Using Figure 2.1, determine
    a. the number of grams of KNO_3 that will dissolve in 100 g of H_2O at 100°C.

    _____ g KNO_3

    b. the number of grams of water required to dissolve 20 g of KNO_3 at 100°C. (Hint: your
    answer to Part a gives you the needed conversion factor for g KNO_3 to g H_2O.)

    _____ g H_2O

    c. the number of grams of water required to dissolve 2.0 g $CuSO_4 \cdot 5\ H_2O$ at 100°C.

    _____ g H_2O

    d. the number of grams of water required to dissolve a mixture containing 20 g KNO_3 and
    2.0 g $CuSO_4 \cdot 5\ H_2O$, assuming that the solubility of one substance is not affected by the
    presence of another.

    _____ g H_2O

2. To the solution in Problem 1d at 100°C, 5 g of water are added, and the solution is cooled to
    0°C.
    a. How much KNO_3 remains in solution?

    _____ g KNO_3

    b. How much KNO_3 crystallizes out?

    _____ g KNO_3

    c. How much $CuSO_4 \cdot 5\ H_2O$ crystallizes out?

    _____ g $CuSO_4 \cdot 5\ H_2O$

    d. What per cent of the KNO_3 in the sample is recovered?

    _____ %   **13**

EXPERIMENT

3 • Resolution of Matter into Pure Substances, II. Paper Chromatography

The fact that different substances have different solubilities in a given solvent can be used in several ways to effect a separation of substances from mixtures in which they are present. We have seen in a previous experiment how fractional crystallization allows us to obtain pure substances by relatively simple procedures based on solubility properties. Another widely used resolution technique, which also depends on solubility differences, is chromatography.

In the chromatographic experiment a mixture is deposited on some solid adsorbing substance, which might consist of a strip of filter paper, a thin layer of silica gel on a piece of glass, some finely divided charcoal packed loosely in a glass tube, or even some microscopic glass beads coated very thinly with a suitable adsorbing substance and contained in a piece of copper tubing.

The components of a mixture are adsorbed on the solid to varying degrees, depending on the nature of the component, the nature of the adsorbent, and the temperature. A solvent is then caused to flow through the adsorbent solid under applied or gravitational pressure or by the capillary effect. As the solvent passes the deposited sample, the various components tend, to varying extents, to be dissolved and swept along the solid. The rate at which a component will move along the solid depends on its relative tendency to be dissolved in the solvent and adsorbed on the solid. The net effect is that, as the solvent passes slowly through the solid, the components separate from each other and move along as rather diffuse zones. With the proper choice of solvent and adsorbent, it is possible to resolve many complex mixtures by this procedure. If necessary, we can usually recover a given component by identifying the position of the zone containing the component, removing that part of the solid from the system, and eluting the desired component with a suitable good solvent.

The name given to a particular kind of chromatography depends upon the manner in which the experiment is conducted. Thus, we have column, thin-layer, paper, and vapor chromatography, all in very common use. Chromatography in its many possible variations offers the chemist one of the best methods, if not the best method, for resolving a mixture into pure substances, regardless of whether that mixture consists of a gas, a volatile liquid, or a group of nonvolatile, relatively unstable, complex organic compounds.

In this experiment we will use paper chromatography to separate a mixture of metallic ions in solution. A sample containing a few micrograms of ions is applied as a spot near one edge of a piece of filter paper. That edge is immersed in a solvent, with the paper held vertically. As the solvent rises up the paper by capillary action, it will carry the metallic ions along with it to a degree that depends upon the relative tendency of each ion to dissolve in the solvent and adsorb on the paper. Because the ions differ in their properties, they move at different rates and become separated on the paper. The position of each ion during the experiment can be recognized if the ion is colored, as some of them are. At the end of the experiment their positions are established more clearly by treating the paper with a staining reagent which reacts with each ion to produce a colored product. By observing the position and color of the spot produced by each ion, and the positions of the spots produced by an unknown containing some of those ions, you can readily determine the ions present in the unknown.

It is possible to describe the position of spots such as those you will be observing in terms of a quantity called the R_f value. In the experiment the solvent rises a certain distance, say L centimeters. At the same time a given component will usually rise a smaller distance, say D centimeters. The ratio of D/L is called the R_f value for that component:

$$R_f = \frac{D}{L} = \frac{\text{distance component moves}}{\text{distance solvent moves}} \qquad (1)$$

The R_f value is a characteristic property of a given component in a chromatography experiment conducted under particular conditions. It does not depend upon concentration or upon the other components present. Hence it can be reported in the literature and used by other researchers doing similar analyses. In the experiment you will be doing, you will be asked to calculate the R_f values for each of the cations studied.

EXPERIMENTAL PROCEDURE

From the stockroom obtain an unknown and a piece of filter paper about 19 cm long and 11 cm wide. Along the 19-cm edge, draw a pencil line about 1 cm from that edge. Starting 1.5 cm from the end of the line, mark the line at 2-cm intervals. Label the segments of the line as shown in Figure 3.1, with the formulas of the ions to be studied and the known and unknown mixtures.

Put two or three drops of 0.1 M solutions of the following compounds in small micro test tubes, one solution to a tube:

$$AgNO_3 \quad Co(NO_3)_2 \quad Cu(NO_3)_2 \quad Fe(NO_3)_3 \quad Hg(NO_3)_2$$

In solution these substances exist as ions. The metallic cations are Ag^+, Co^{2+}, Cu^{2+}, Fe^{3+}, and Hg^{2+} respectively. One drop of each solution contains about 50 micrograms of cation. Into a sixth micro test tube put two drops of each of the five solutions; swirl until the solutions are well mixed. This mixture will be our known, since we know it contains all of the cations.

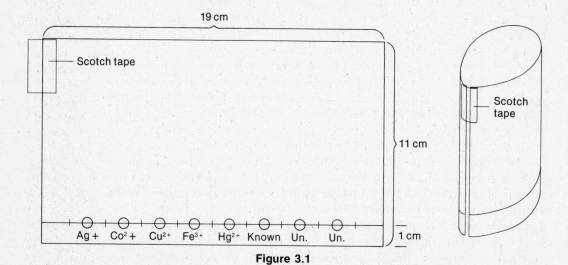

Figure 3.1

Your instructor will furnish you with a fine capillary tube, which will serve as an applicator. Test the application procedure by dipping the applicator into one of the colored solutions and touching it momentarily to a round piece of filter paper. The liquid from the applicator should form a spot no larger than 8 mm in diameter. Practice making spots until you can reproduce the spot size each time.

Clean the applicator by dipping it about 1 cm into distilled water and then touching the round filter paper to remove the liquid. Continue contact until all the liquid in the tube is gone. Repeat the cleaning procedure one more time. Dip the applicator into one of the cation solutions and put a spot on the line on the rectangular filter paper in the region labeled for that cation. Clean the applicator twice, and repeat the procedure with another solution. Continue this approach until you have put a spot for each of the five cations and the known and unknown on the paper, cleaning the applicator between solutions. Dry the paper by moving it in the air or holding it briefly in front of a hair dryer or heat lamp (low setting). Apply the known and unknown twice more to the same spots; the known and unknown are less concentrated than the cation solutions, so this procedure will increase the amount of each ion in the spots. Make sure that you dry the spots between applications, since otherwise they will get larger. Don't heat the paper more than necessary to just dry the spots.

Draw about 15 ml of eluting solution from the supply on the reagent shelf. This solution is made by mixing a solution of HCl, hydrochloric acid, with ethanol and butanol, which are organic solvents. Pour the eluting solution into a 600-ml beaker and cover with a watch glass.

Check to make sure that the spots on the filter paper are all dry. Place a 4- to 5-cm length of Scotch tape along the upper end of the left edge of the paper, as shown in Figure 3.1, so that about half of the tape is on the paper. Form the paper into a cylinder by attaching the tape to the other edge, in such a way that the edges are parallel but do not overlap. When you are finished, the pencil line at the bottom of the cylinder should form a circle, approximately anyway, and the two edges of the paper should not quite touch. Stand the cylinder up on the lab bench to check that such is the case and readjust the tape if necessary. *Do not* tape the lower edges of the paper together.

Place the cylinder in the eluting solution in the 600-ml beaker, with the sample spots down near the liquid surface. The paper should not touch the wall of the beaker. Cover the beaker with the watch glass. The solvent will gradually rise by capillary action up the filter paper, carrying along the cations at different rates. After the process has gone on for a few minutes, you should be able to see colored spots on the paper, showing the positions of some of the cations.

While the experiment is proceeding, you can test the effect of the staining reagents on the different cations. Put an 8-mm spot of each of the cation solutions on a clean piece of round filter paper, labelling each spot and cleaning the applicator between solutions. Dry the spots as before. Some of them will have a little color; record those colors on the Data sheet. Then go to a hood, and hold the paper over an open bottle of 15 M NH_3, ammonia. The ammonia will react with some of the cations, forming colored products. Note any colors that develop and record them. Then put the paper in a large beaker, in the hood, in which there is a solution of $(NH_4)_2S$, ammonium sulfide. Don't touch the paper to the liquid. The vapor above the solution contains H_2S, which will react with all of the cations and produce colored sulfides. Record the colors of the sulfides. Considering that each spot contains less than 50 micrograms of cation, the tests are quite definitive.

When the eluting solution has risen to about 2 cm from the top of the filter paper (it will take about 75 minutes), remove the cylinder from the beaker and take off the tape. Draw a pencil line along the solvent front. Dry the paper in air, or with gentle heat, until it is just about, but not quite, dry. Note any cations that must be present in your unknown by virtue of your being able to see their colors. Then hold the paper above the bottle of 15 M NH_3, and report any cations that are revealed by the color of their spots. Finally hold the paper in the beaker containing $(NH_4)_2S$, and report those cations whose sulfides you can see. Any cations you identified in your unknown before staining, or with NH_3, should be observed as sulfides.

Measure the distance from the straight line on which you applied the spots to the solvent front, which is distance L in Equation 1. Then measure the distance from the pencil line to the center of the spot made by each of the cations, when pure and in the known; this is distance D. Calculate R_f value for each cation. Then calculate R_f values for the cations in the unknown. How do the R_f values compare?

Name _____ Section _____

DATA AND CALCULATIONS: Resolution of Matter into Pure Substances, II.
                           Paper Chromatography

		Ag^+	Co^{2+}	Cu^{2+}	Fe^{3+}	Hg^{2+}
Colors (if observed)	Dry	_____	_____	_____	_____	_____
	In NH_3	_____	_____	_____	_____	_____
	In $(NH_4)_2S$	_____	_____	_____	_____	_____
Distance solvent moved (L)		_____	_____	_____	_____	_____
Distance cation moved (D)		_____	_____	_____	_____	_____
R_f		_____	_____	_____	_____	_____

Known Mixture

	Ag^+	Co^{2+}	Cu^{2+}	Fe^{3+}	Hg^{2+}
Distance solvent moved	_____	_____	_____	_____	_____
Distance cation moved	_____	_____	_____	_____	_____
R_f	_____	_____	_____	_____	_____

Unknown Mixture

	Ag^+	Co^{2+}	Cu^{2+}	Fe^{3+}
Cations identified when dry	_____	_____	_____	_____
when in NH_3	_____	_____	_____	_____
when in $(NH_4)_2S$	_____	_____	_____	_____
Distance solvent moved	_____	_____	_____	_____
Distance cation moved	_____	_____	_____	_____
R_f	_____	_____	_____	_____
Composition of unknown	_____	_____	_____	_____

Unknown No. _____

ADVANCE STUDY ASSIGNMENT: Resolution of Matter into Pure Substances, II.
                            Paper Chromatography.

1.   A student chromatographs a mixture, and after developing the spots with a suitable reagent he observes the following:

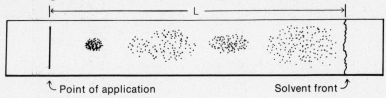

What are the R_f values?

2.   Explain, in your own words, why samples can often be separated into their components by chromatography.

3.   The solvent moves 3 cm in about 10 minutes. Why shouldn't the experiment be stopped at that time instead of waiting 75 minutes for the solvent to move 10 cm?

4.   In this experiment it takes about 10 microliters of solution to produce a spot 1 cm in diameter. If the $Cu(NO_3)_2$ solution contains about 6 g Cu^{2+} per liter, how many micrograms of Cu^{2+} ion are there in one spot?

_____ micrograms

21

EXPERIMENT

4 • Determination of a Chemical Formula

When a substance A reacts with another substance B to form a third substance C, the equation for the chemical reaction can be written as

$$aA + bB \rightarrow cC \tag{1}$$

The substances A and B may be atoms, molecules, or ions in aqueous solution. The numbers a, b, and c are small integers and denote the relative numbers of particles involved in the reaction. Since a mole of any substance contains the same number of particles, be they atoms, molecules, or ions, the numbers a, b, and c also indicate the numbers of moles of A and B that react to form C. There are many reactions that conform to Equation 1, including the following examples:

$$2 \text{ H}_2(g) + \text{O}_2(g) \rightarrow 2 \text{ H}_2\text{O}(l) \tag{2}$$

$$3 \text{ Ca}^{2+}(aq) + 2 \text{ PO}_4^{3-}(aq) \rightarrow \text{Ca}_3(\text{PO}_4)_2(s) \tag{3}$$

Reaction 3 would occur when a solution containing Ca^{2+} ions was mixed with one containing phosphate, PO_4^{3-}, ions. Since the reaction goes essentially to completion there will ordinarily be an excess of one of the reacting ions in the mixture, since the other one will be all used up. If, for example, one slowly added a solution of PO_4^{3-} ions to one containing Ca^{2+} ions, the phosphate ions would react to form $\text{Ca}_3(\text{PO}_4)_2$ as fast as we added them, so there would, early on, be very little PO_4^{3-} left in solution and an excess of Ca^{2+} ions. As we continued to add PO_4^{3-} ions, more precipitate of $\text{Ca}_3(\text{PO}_4)_2$ would form, using up Ca^{2+} ion, until finally all of the Ca^{2+} initially present would be reacted. After that, further addition of phosphate ion (now in excess) would raise the concentration of that ion, while the concentration of Ca^{2+} ion would remain about zero.

If in carrying out Reaction 3 by the method we describe, we could stop when all of the Ca^{2+} had been converted to $\text{Ca}_3(\text{PO}_4)_2$, we could verify the formula for $\text{Ca}_3(\text{PO}_4)_2$ by noting the relative numbers of moles of Ca^{2+} initially in the solution and of PO_4^{3-} that were added. In this case we would find that we needed 2 moles of PO_4^{3-} for every 3 moles of Ca^{2+} in the original solution. This would tell us that the formula of calcium phosphate has to be $\text{Ca}_3(\text{PO}_4)_2$. In this experiment we will find the chemical formula of an insoluble salt containing metallic cations and chromate, CrO_4^{2-}, anions by using this approach.

In our procedure we will first weigh out a sample of a soluble salt containing a cation that forms an insoluble chromate. An example of such a salt is lead nitrate, $\text{Pb}(\text{NO}_3)_2$. This salt will serve as our source of metallic cations. Given the mass of the sample, and its formula, we can calculate the number of moles of salt in the sample and, more important, the number of moles of metallic cation it contains. Assuming, say, that we had $\text{Pb}(\text{NO}_3)_2$ in our sample, and that it weighed 0.4518 grams, we would proceed as follows:

Molar mass $\text{Pb}(\text{NO}_3)_2$ = (atomic mass Pb + 2 × atomic mass N + 6 × atomic mass O) grams
= (207.2 + 2 × 14.0067 + 6 × 15.9994) grams = 331.2 grams

Number of moles of $\text{Pb}(\text{NO}_3)_2$ = $\dfrac{\text{mass of sample}}{\text{molar mass Pb}(\text{NO}_3)_2}$ = $\dfrac{0.4518 \text{ g}}{331.2 \text{ g/mole}}$ = 1.364×10^{-3} moles

Number of moles of Pb^{2+} = number of moles of $\text{Pb}(\text{NO}_3)_2$ = 1.364×10^{-3} moles

Having weighed our sample, we will dissolve it in 20 ml of water, forming just about 20 ml of solution. In the solution the $\text{Pb}(\text{NO}_3)_2$ will break up, or ionize, completely into Pb^{2+} and NO_3^{-}

ions. We can calculate the number of moles of Pb^{2+} ions present in one milliliter of solution very easily:

$$\text{Number of moles of } Pb^{2+} \text{ per ml of solution} = \frac{\text{Number of moles of } Pb^{2+}}{\text{Volume of solution in ml}}$$

$$= \frac{1.364 \times 10^{-3} \text{ moles}}{20.0 \text{ ml}} = 6.82 \times 10^{-5} \frac{\text{moles/ml}}{}$$

We add exactly 1 ml of the solution we have prepared to each of 6 small test tubes, numbered 1 to 6.

To test tube No. 1 we then add 1 ml 0.020 M K_2CrO_4. This solution contains 0.020 moles K_2CrO_4 per liter, and since this salt, like all salts, is ionized in solution, it also contains 0.020 moles CrO_4^{2-} per liter, or 2.0×10^{-5} moles CrO_4^{2-} per milliliter. As soon as the Pb^{2+} and CrO_4^{2-} ions meet, they react to form a yellow precipitate of lead chromate. To test tube No. 2 we add 2 ml 0.020 M K_2CrO_4, to test tube No. 3, 3 ml, and so on through test tube No. 6, to which we add 6 ml of the chromate solution.

In some of the test tubes Pb^{2+} ion is in excess, since not enough CrO_4^{2-} ion was added to precipitate all of the cation. In some of the other tubes CrO_4^{2-} ion will be in excess, since more than enough was added to precipitate all of the Pb^{2+} ion. We can determine which ion is in excess in each tube by centrifuging to bring the solid precipitate to the bottom of the tube. The strong yellow color of chromate ion will be apparent in the solutions in those tubes where CrO_4^{2-} is in excess. Where Pb^{2+} is in excess, the solution will be essentially colorless.

If we did this experiment with the sample of $Pb(NO_3)_2$ we used in our example we would find that Mixtures No. 1, 2, and 3 were colorless after centrifuging and that Mixtures No. 4 through 6 were yellow. This allows us to say that in Mixture No. 3 Pb^{2+} was in excess, while in Mixture No. 4 CrO_4^{2-} was in excess. Using the mixtures in those two tubes we can calculate, within limits, the formula for lead chromate. We proceed as follows:

In Mixture No. 3: No. moles $Pb^{2+} = 6.82 \times 10^{-5}$ moles
   No. moles $CrO_4^{2-} = 3.0$ ml $\times 2.0 \times 10^{-5}$ moles/ml $= 6.0 \times 10^{-5}$ moles

Mole ratio, $CrO_4^{2-} : Pb^{2+} = 6.0 \times 10^{-5}$ moles/6.82×10^{-5} moles $= 0.88 : 1.00$

In Mixture No. 4: No. moles $Pb^{2+} = 6.82 \times 10^{-5}$ moles
   No. moles $CrO_4^{2-} = 4.0$ ml $\times 2.0 \times 10^{-5}$ moles/ml $= 8.0 \times 10^{-5}$ moles

Mole ratio, $CrO_4^{2-} : Pb^{2+} = 8.0 \times 10^{-5}$ moles/6.82×10^{-5} moles $= 1.2 : 1.00$

If, in Mixtures No. 3 and 4, we assume that *all* the Pb^{2+} and CrO_4^{2-} are present as solid lead chromate, the formulas we would obtain for the compound would be $Pb(CrO_4)_{0.88}$ in Mixture No. 3 and $Pb(CrO_4)_{1.2}$ in Mixture No. 4. The true formula must lie somewhere between these values, since in Mixture No. 3 Pb^{2+} ion is in excess and in Mixture No. 4 CrO_4^{2-} is in excess. Since the mole ratio for $Pb^{2+} : CrO_4^{2-}$ is expected to be a ratio of small integers, we would guess that the ratio might well be 1:1, and that the associated formula for lead chromate would be $PbCrO_4$.

In the optional part of this experiment we will "fine-tune" the amount of CrO_4^{2-} solution we add, getting closer to the situation in which we add just the right amount to precipitate all of the metallic cation. This will allow us to limit the formula to a narrower range than we get with the first set of mixtures.

EXPERIMENTAL PROCEDURE   SEE WARNING AT END OF EXPERIMENT

Part I

From the stockroom obtain a sample of a soluble salt. You will be given its chemical formula at that time.

On an analytical balance, weigh a clean, dry 50-ml beaker to 0.001 g. Add your sample to the beaker and weigh once again to the same precision.

Using your graduated cylinder, add 20.0 ml of distilled water to the sample in the beaker. Stir with your stirring rod until the sample is dissolved. Then stir for 2 full minutes to ensure that the solution is thoroughly mixed. Pour some 0.020 M K_2CrO_4 from the stock solution into another clean, dry 50-ml beaker until the beaker is about $\frac{2}{3}$ full. Use this solution as your source of chromate ion.

Set up a hot-water bath, consisting of a 250-ml beaker $\frac{2}{3}$ full of water supported on a wire gauze on an iron ring fastened to a ring stand. Begin heating the water with a Bunsen burner.

Number six small test tubes from 1 to 6 and put them in your test tube rack. The test tubes should be clean but not necessarily dry. Using a graduated 5-ml pipet, add exactly 1 ml of your salt solution to each of the test tubes. Your instructor will demonstrate how to use the pipet properly. Be careful when measuring out these small volumes. Then, using a 10-ml graduated pipet, add 1 ml 0.020 M K_2CrO_4 to test tube No. 1, 2 ml to test tube No. 2, 3 ml to test tube No. 3, and 4 ml to test tube No. 4. Refill the pipet and put 5 ml and 6 ml of the chromate solution in test tubes No. 5 and 6, respectively. To the first five test tubes add water from your wash bottle until the volume in each tube is about that in test tube No. 6. The volumes of reagents in each test are summarized in Table 4.1.

TABLE 4.1 COMPOSITION OF REACTION MIXTURES FOR CHROMATE PRECIPITATIONS

Test tube no.	1	2	3	4	5	6
ml salt solution	1	1	1	1	1	1
ml 0.020 M K_2CrO_4	1	2	3	4	5	6
ml water (approx.)	5	4	3	2	1	0
Mixture no.	1	2	3	4	5	6

Shake each tube for at least 30 seconds, using a cork stopper and wiping the cork on a piece of paper towel between tubes. Then put all of the tubes in the hot-water bath, which should be simmering by now. Let the tubes stay in the water bath for about five minutes to promote formation of large crystals of the chromate precipitate. Keep the water bath at the simmering point by judicious adjustment of your Bunsen burner. Remove the tubes from the bath and centrifuge them in batches of two, four, or six tubes, depending on the centrifuge capacity. Then arrange the tubes in order of increasing number from left to right in your test tube rack.

If all goes well, you will find that the solutions above the first few mixtures, with the lower numbers, will be essentially colorless, indicating that the cation is in excess, while the rest of the mixtures will produce yellow solutions, owing to the presence of excess chromate ion. There should be two adjacent tubes in the set of six, one containing a colorless solution and a slight excess of cation and the other yellow with a slight excess of chromate ion. Record the numbers of those two tubes.

Since the two tubes whose numbers you recorded contain mixtures that differ by 1 ml in the amount of chromate solution used to prepare them, we know, within 1 ml, the volume of chromate solution needed to precipitate all of the cation in 1 ml of salt solution. We will call the test tube with the colorless solution Tube A and the one with the yellow solution Tube B.

Part II (Optional)

In this part of the experiment we will attempt to determine to within about 0.2 ml the volume of chromate solution needed to precipitate the cation in 1 ml of the salt solution you prepared. We will do this by adding small but increasing volumes of the chromate solution to six mixtures, all made up identical to the one in Tube A, and detecting the point at which change in solution color occurs, in much the same way as in Part I.

Pour the six mixtures you prepared in Part I into the waste crock on your lab bench. Clean out the test tubes with your test tube brush, rinse the tubes well with distilled water, and put them back in the test tube rack in order of increasing number.

Carefully pipet 1 ml of your salt solution into each of the six tubes, using the 5-ml graduated pipet. Then pipet the same number of ml of the 0.020 M K_2CrO_4 solution used to make the mixture in Tube A into each of those six tubes. Essentially you should make six replicas of the mixture that was in Tube A.

Fill a medicine dropper with distilled water from a small beaker, and add the water, drop by drop, to your 10-ml graduated cylinder. Count the number of drops needed to deliver the first ml, the second, and the third, until you are sure you know how many drops it takes to make 1 ml, within a drop or two.

To test tubes No. 1 to 6 we want to add chromate solution in slowly increasing volumes, starting with roughly 0.2 ml in tube No. 1 and ending with about 1 ml in tube No. 6. So, take the number of drops per ml, as delivered by your dropper, and divide by six. If you don't get an integer, take the next larger whole number of drops. (If you have 15 drops per milliliter, you would take 3 drops; if you have 18 drops per ml, you would also take 3 drops.)

Rinse your medicine dropper with the 0.020 M K_2CrO_4 solution in your beaker. Then add the number of drops you just calculated to test tube No. 1, twice that number to test tube No. 2, three times that number to test tube No. 3, and so on, finally adding six times that number to test tube No. 6. In this way, we will get six different mixtures, some with excess cation and some with excess chromate ion, as in Part I. Add distilled water from your wash bottle to each tube to bring the volume up to within about 2 cm of the top of the tube.

Now proceed as you did in Part I, shaking each tube after stoppering it, heating the tubes in the water bath, and centrifuging to bring down the chromate precipitate. Then arrange the tubes in the test-tube rack in order of increasing number.

Once again, there should be a series of tubes with lower numbers above which the solution is colorless, followed by a series with the large numbers for which the solution is yellow. Since now the difference in composition is not as great as in Part I, the changes in color will not be so great, but you should still be able to select two adjacent tubes, one with essentially a colorless solution and the next with a yellow one. Record the numbers of those two tubes, which we will now call Tubes C and D, respectively.

CAUTION: The solids used in this experiment are toxic. Avoid contact with the crystals or with their solutions. Wash your hands when you finish the experiment.

Name_____ Section_____

DATA AND CALCULATIONS: Determination of a Chemical Formula

Part I

Formula of salt_____   Molar mass of salt_____ g

Mass of empty beaker _____ g

Mass of beaker plus salt_____ g

Mass of salt   _____ g

Number of moles of salt in sample   _____ moles

Number of moles of salt in 1 ml of solution   _____ moles

Number of moles of cation in 1 ml of solution   _____ moles

Number of test tube containing colorless solution (Tube A)   _____

Number of adjacent test tube containing yellow solution (Tube B)   _____

	Tube A	Tube B
No. of moles of cation added	_____ moles	_____ moles
Volume of chromate solution added	_____ ml	_____ ml
No. of moles of CrO_4^{2-} added	_____ moles	_____ moles
Mole ratio, CrO_4^{2-} : cation	_____	_____

Formula of chromate in Tube A, assuming complete precipitation   _____

Formula of chromate in Tube B, assuming complete precipitation   _____

Probable formula of your chromate salt   _____

Continued on following page

Part II (Optional)

Number of drops per ml from dropper _____ _____ _____   Avg._____

Volume of one drop_____ml

Number of drops of 0.020 M K_2CrO_4 added to Test Tube No. 1   _____drops

Number of test tube containing colorless solution (Tube C)   _____

Number of adjacent test tube containing yellow solution (Tube D)   _____

	Tube C	Tube D
No. of moles of cation added	_____ moles	_____moles
Pipetted volume of chromate solution	_____ml	_____ml
No. drops chromate solution added	_____drops	_____drops
Volume of drops of chromate solution	_____ml	_____ml
Total volume chromate solution added	_____ml	_____ml
No. of moles of $CrO_4{}^{2-}$ added (2 sig. fig.)	_____ moles	_____moles
Mole ratio, $CrO_4{}^{2-}$:cation	_____	_____

Formula of chromate in Tube C, assuming complete precipitation   _____

Formula of chromate in Tube D, assuming complete precipitation   _____

Formula of chromate_____

ADVANCE STUDY ASSIGNMENT: Determination of a Chemical Formula

1.   A student performing this experiment was given mercury (II) nitrate, $Hg(NO_3)_2 \cdot H_2O$ as her salt. She weighed the salt in a beaker on the analytical balance and found its mass to be 0.5212 g.

     a. What was the molar mass of her sample?

$$\text{Molar mass} = (AM_{Hg} + 2\ AM_N + 6\ AM_O + 2\ AM_H + AM_O)\ grams$$

     where AM is atomic mass

                                                            _____ grams

     b. How many moles of salt were there in her sample? How many moles of Hg^{2+}?

$$\text{No. moles} = \frac{\text{mass of sample}}{\text{molar mass of sample}}$$

                          _____ moles; _____ moles Hg^{2+}

2.   The student dissolved the sample in 20 ml of water and then mixed the solution thoroughly. How many moles of Hg^{2+} were there in 1 ml of the solution?

$$\text{No. moles } Hg^{2+} \text{ per ml} = \frac{\text{No. moles } Hg^{2+}}{\text{Volume of soln in ml}}$$

                                          _____ moles/ml

3.   On completing Part I of the experiment, the student found that the solutions in test tubes No. 1 through 3 were colorless, while those in test tubes No. 4 through 6 were yellow. She selected tubes No. 3 and 4 as Tube A and Tube B.

     a. How had she made up the mixture in Tube A? Tube B?

     b. How many moles of Hg^{2+} were there in Tube A? Tube B?

                     _____ moles Hg^{2+}; _____ moles Hg^{2+}

     c. How many moles of CrO_4^{2-} were there in Tube A? Tube B?

$$\text{No. moles } CrO_4^{2-} = \text{No. ml chromate soln.} \times 2.0 \times 10^{-5} \text{ moles/ml}$$

                     _____ moles CrO_4^{2-}; _____ moles CrO_4^{2-}

Continued on following page   **29**

d. What was the mole ratio, $CrO_4^{2-}:Hg^{2+}$, in Tube A? in Tube B?

$$\text{Mole ratio} = \frac{\text{no. moles } CrO_4^{2-}}{\text{no. moles } Hg^{2+}}$$ _____ in Tube A: _____ in Tube B

e. If all of the Hg^{2+} and CrO_4^{2-} ions were in the form of precipitated mercury (II) chromate in Tube A, what would be the formula of that chromate? _____

f. If all of the Hg^{2+} and CrO_4^{2-} in Tube B were in the form of precipitated solid chromate, what would its formula be? _____

g. What is the probable formula of the mercury (II) chromate?   _____

4. In Part II the student found that her medicine dropper delivered 17 drops/ml. She prepared six tubes, all made up with the mixture she had used in Tube A. To tubes No. 1 through 6 she added 3, 6, 9, 12, 15, and 18 drops, respectively. After heating and centrifuging, she found that the solutions in tubes No. 1 through 4 were colorless and those in tubes No. 5 and 6 were yellow. She therefore chose tubes No. 4 and 5 as Tube C and Tube D, respectively.

a. How many moles of Hg^{2+} were there in Tube C? in Tube D?

_____ moles Hg^{2+}; _____ moles Hg^{2+}

b. How many ml of chromate solution had been pipetted into Tube C? into Tube D?

_____ ml; _____ ml

c. How many drops of chromate solution had been added to Tube C? to Tube D?

_____ drops; _____ drops

d. What was the volume of a drop?   _____ ml

e. What was the volume of the drops added to Tube C? to Tube D?

_____ ml to Tube C: _____ ml to Tube D

f. What was the total volume of chromate solution added to Tube C? to Tube D?

Total volume = sum of parts b and e

_____ ml to Tube C: _____ ml to Tube D

g. How many moles of CrO_4^{2-} were added to Tube C? to Tube D?

No. moles = total volume added in ml $\times 2.00 \times 10^{-5}$ moles/ml

(2 significant figures)   _____ moles CrO_4^{2-}; _____ moles CrO_4^{2-}

h. What was the mole ratio, $CrO_4^{2-}:Hg^{2+}$, in Tube C? in Tube D? _____; _____

i. What is the formula of mercury (II) chromate?   _____

EXPERIMENT

5 • Law of Multiple Proportions

Many elements form more than one compound with oxygen or other nonmetals. For example, iron forms three oxides, with the formulas FeO, Fe_2O_3, and Fe_3O_4. These oxides, like any set of binary compounds containing the same two elements, can be used to illustrate the Law of Multiple Proportions. If we start with a mole of Fe, the numbers of moles of O combining with that amount of iron are 1, 3/2, and 4/3, respectively; all these numbers are either integers or simple fractions, as required by the Law. By analyzing a set of compounds like the oxides of iron we can find their formulas and check the validity of the Law of Multiple Proportions.

In this experiment we will be studying some binary compounds containing copper and bromine. You will be furnished a sample of one of these copper bromides (compound A). On heating a weighed sample of compound A you will find that it breaks down to another copper bromide (compound B), liberating bromine, Br_2, in the process. The reaction might be described as

$$CuBr_x(s) \xrightarrow{\text{heat}} CuBr_y + \frac{(x-y)}{2} Br_2(g) \tag{1}$$
$$\quad\text{A} \qquad\qquad \text{B}$$

We will then convert compound B to an oxide of copper (compound C) by treating it with nitric acid and then heating it. The reactions are fairly complex but the net effect can be represented by the equation:

$$CuBr_y(s) + \frac{z}{2} O_2(g) \rightarrow CuO_z(s) + \frac{y}{2} Br_2(g) \tag{2}$$

The copper oxide is then reduced to copper by heating in the presence of natural gas. By weighing the sample at the end of each step in this series of reactions, we can find the masses of compounds A, B, and C, along with the mass of the copper they contain. From these data, and the atomic masses of copper, bromine, and oxygen, we can find the formulas of the two bromides and the oxide and determine whether the Law of Multiple Proportions applies to copper-bromine compounds.

EXPERIMENTAL PROCEDURE

WEAR YOUR SAFETY GLASSES WHILE PERFORMING THIS EXPERIMENT

Place about one gram of the copper bromide (A) in a large, accurately weighed test tube. Weigh the test tube and its contents to the nearest 0.001 g. Clamp the test tube so that it is slightly inclined, and incorporate it into an apparatus such as that shown in Figure 5.1. The bent glass tube should extend about an inch past the end of the rubber stopper, and should go to within about an inch of the surface of the solution in the 500-ml Florence flask. *Do not* let the end of the tube dip into the liquid; if it does, liquid will back up into the test tube during the experiment and cause trouble.

Heat the bromide sample, first gently and then rather strongly, with the burner flame. The sample will decompose and bromine, Br_2, will be evolved and absorbed in the solution. *Bromine is a very reactive and very poisonous substance. Do not inhale the vapor or touch the dark liquid.* Continue heating until bromine is no longer produced. (Do not heat the test tube to redness during this step, since at high temperatures the initial decomposition product may be further decomposed to copper.) Let the test tube cool, and remove it from the apparatus. If there is any bromine

31

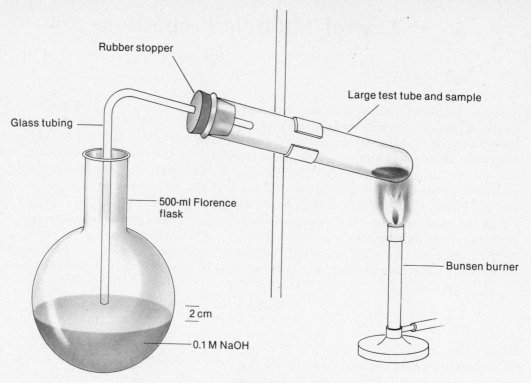

Figure 5.1

condensed on the walls of the tube, remove it by warming the test tube gently *in the hood* with a Bunsen flame. Allow the tube to cool and then weigh it on the balance.

Add 2 cm³ of 15 M HNO₃ (**CAUTION: CAUSTIC REAGENT**) to the test tube and reassemble the apparatus as before. Heat the test tube gently and then strongly for a few minutes to produce black copper oxide (compound C). Weigh the test tube and its contents when cool.

In the last part of the experiment we will reduce the copper oxide to metallic copper with zinc and hydrochloric acid. Clamp the tube in a vertical position. Add 5 ml 6 M HCl, hydrochloric acid. The copper oxide should dissolve readily, forming a dark green solution. Heat the tube *gently*, if necessary, to complete the dissolution. Check to see that any copper oxide on the upper walls of the tube has been dissolved. Add 5 ml distilled water.

Weigh out about 0.5 grams of granular zinc on the top-loading balance. Slowly add the zinc to the test tube, over a period of a few minutes. The zinc will produce hydrogen gas as it comes in contact with the acid, and the mixture will reduce any copper present to metallic copper. As the reaction proceeds, the color of the mixture will lighten, becoming yellow, and eventually colorless.

When you have added all the zinc, heat the tube *gently* to speed up the reaction. *Do not* boil the mixture. If foam carries solid up the tube walls, wash it down with a *little* water from your wash bottle. When the evolution of hydrogen has ceased, and the solution is colorless, stop heating and let the copper settle out as the tube cools. (If the solution is not colorless, add a little more zinc, and heat gently until the color disappears and gas is no longer produced.) Carefully pour off the liquid, leaving the copper in the tube. Add 10 ml distilled water and 1 ml 6 M HCl, swirl to mix the reagents, and heat for five more minutes, without boiling. Let the solid settle, and pour off the liquid. Wash the solid copper four times with 10-ml portions of distilled water, decanting the water, but not the copper, each time.

Clamp the test tube at a 45° angle and heat the copper gently to drive off the remaining water. As the water vaporizes, the copper particles may "pop." Heat the upper walls of the tube to vaporize any water that condenses. When the copper, and the tube, are dry, let the tube cool to room temperature. Weigh the test tube and the copper it contains on the analytical balance.

DATA: Law of Multiple Proportions

Mass of empty test tube   _____ g

Mass of test tube plus copper bromide (A)   _____ g

Mass of test tube plus copper bromide (B)   _____ g

Mass of test tube plus copper oxide (C)   _____ g

Mass of test tube plus copper   _____ g

Mass of copper bromide (A)   _____ g

Mass of copper bromide (B)   _____ g

Mass of copper oxide (C)   _____ g

Mass of copper   _____ g

CALCULATIONS

1. Using the data you obtained, calculate the amounts of copper and bromine present in your samples of copper bromide (A) and copper bromide (B).

Compound A: _____ g Cu and _____ g Br

Compound B: _____ g Cu and _____ g Br

2. From the results obtained in Part 1, calculate the number of grams of bromine, Br, that would combine with *one mole* (63.54 g) of Cu in compounds A and B.

Compound A: 63.54 g Cu $\simeq$ _____ g Br

Compound B: 63.54 g Cu $\simeq$ _____ g Br

Continued on following page

3. Using your results from Part 2, calculate the number of moles of Br that combine with one mole of Cu in the two copper bromide compounds (atomic mass Br = 79.90).

Compound A: 1.000 mol Cu $\simeq$ _____ mol Br

Compound B: 1.000 mol Cu $\simeq$ _____ mol Br

4. What is the simplest formula of compound A? compound B?   _____(A)

_____(B)

5. Show how the Law of Multiple Proportions relates to your results in Parts 1, 2, 3, and 4.

6. Find the simplest formula of the copper oxide, compound C, that you prepared. Use the same general procedure as in finding the formulas of compounds A and B in Parts 1, 2, 3, and 4.

Formula of compound C _____

ADVANCE STUDY ASSIGNMENT: Law of Multiple Proportions

Lead forms two chlorides. One of them is an oily yellow liquid (compound A) and the other is a white solid (compound B). A chemist added a sample of compound A to a test tube weighing 31.867 g. On reweighing the test tube and its contents he found the mass was 33.356 g.

1. How many grams did the sample of compound A weigh?

_____ g

On adding a little water to the test tube, he found that $Cl_2(g)$ was liberated (he was working in a hood, as he should have been), and compound A was converted to compound B. He drove off the water by heating the tube, and found that the mass of the tube and compound B was 33.050 g.

2. What was the mass of the sample of compound B?

_____ g

The chemist then heated compound B in the presence of hydrogen, reducing the lead in the compound to metallic form and driving off the chlorine in the compound as $HCl(g)$. He weighed the test tube and the lead, obtaining a mass of 32.751 g.

3. How many grams of lead were in the tube?

_____ g Pb

The lead in the tube was present in compounds A and B, so they both contained the amount of lead obtained in Part 3. The rest of the mass of each of the samples had to be due to chlorine, Cl.

4. How much chlorine was there in the sample of compound A? compound B?

Compound A _____ g Cl

Compound B _____ g Cl

5. The masses of lead and chlorine obtained in Parts 3 and 4 provide us with the elemental analysis of those compounds. Using those masses, we can say that

in compound A, _____ g Pb $\simeq$ _____ g Cl

in compound B, _____ g Pb $\simeq$ _____ g Cl

Continued on following page

6. Using the equivalences in Part 5 as conversion factors, calculate the number of *grams* of Cl that would combine with *1 mol, 207.2 g,* of Pb in the two compounds.

Compound A: grams Cl = 207.2 g Pb × $\dfrac{\rule{3cm}{0.4pt}\text{g Cl}}{\rule{3cm}{0.4pt}\text{g Pb}}$ =

$\rule{3cm}{0.4pt}$ g Cl

Compound B: grams Cl = 207.2 g Pb × $\dfrac{\rule{3cm}{0.4pt}\text{g Cl}}{\rule{3cm}{0.4pt}\text{g Pb}}$ =

$\rule{3cm}{0.4pt}$ g Cl

7. From the results of Part 6, find the number of *moles* of Cl that combine with *one mole* of Pb in the two lead chloride compounds (atomic mass Cl = 35.45).

Compound A: 1.000 mol Pb combines with $\rule{3cm}{0.4pt}$ mol Cl

Compound B: 1.000 mol Pb combines with $\rule{3cm}{0.4pt}$ mol Cl

8. Given the fact that the atom ratio in a compound must equal the mol ratio, what are the simplest formulas of compounds A and B?

Compound A $\rule{3cm}{0.4pt}$

Compound B $\rule{3cm}{0.4pt}$

9. How can the results of Parts 5, 6, and 7 be used to test the Law of Multiple Proportions?

EXPERIMENT

6 • Water of Hydration

Most solid chemical compounds will contain some water if they have been exposed to the atmosphere for any length of time. In most cases the water is present in very small amounts, and is merely adsorbed on the surface of the crystals. Other solid compounds contain larger amounts of water that is chemically bound in the crystal. These compounds are usually ionic salts. The water that is present in these salts is called water of hydration and is usually bound to the cations in the salt.

The water molecules in a hydrate are removed relatively easily. In most cases simply heating a hydrate to a temperature somewhat above the boiling point of water will drive off the water of hydration. Hydrated copper(II) chloride is typical in this regard; it is converted to anhydrous $CuCl_2$ if heated to about 110°C:

$$CuCl_2 \cdot 2\ H_2O \rightarrow CuCl_2(s) + 2\ H_2O(g) \text{ at } t \geqslant 110°C$$

In the dehydration reaction the crystal structure of the solid will change and the color of the salt may also change. On heating $CuCl_2 \cdot 2\ H_2O$ the green hydrated crystals are converted to a brownish-yellow powder. You may be familiar with hydrated $CoCl_2$, which is sometimes used in inexpensive hygrometers. $CoCl_2 \cdot 6\ H_2O$ is red, $CoCl_2 \cdot 2\ H_2O$ is violet, and $CoCl_2$ is blue.

Some hydrates lose water to the atmosphere upon standing. This process is called efflorescence. The amount of water lost depends upon the amount of water in the air, as measured by its relative humidity. In moist, warm air, $CoCl_2$ is fully hydrated and is red; in dry cold air $CoCl_2$ loses most of its water of hydration and is blue; at intermediate humidities, $CoCl_2$ exists as a dihydrate and is violet.

Some anhydrous ionic compounds will tend to absorb water from the air or other sources so strongly that they can be used to dry liquids or gases. These substances are called desiccants, and are said to be hygroscopic. A few ionic compounds can take up so much water from the air that they dissolve in the water they absorb; sodium hydroxide, $NaOH$, will do this. This process is called deliquescence.

Some compounds evolve water on being heated but are not true hydrates. The water is produced by decomposition of the compound rather than by loss of water of hydration. Organic compounds, particularly carbohydrates, behave this way. Decompositions of this sort are not reversible; adding water to the product will not regenerate the original compound. True hydrates typically undergo reversible dehydration. Adding water to anhydrous $CuCl_2$ will cause formation of $CuCl_2 \cdot 2\ H_2O$ or, if enough water is added, you will get a solution containing hydrated Cu^{2+} ions. All ionic hydrates are soluble in water, and are usually prepared by crystallization from water solution. The amount of bound water may depend upon the way the hydrate is prepared, but in general the number of moles of water per mole of ionic compound is either an integer or a multiple of $\frac{1}{2}$.

In this experiment you will study some of the properties of hydrates. You will identify the hydrates in a group of compounds, observe the reversibility of the hydration reaction, and test some substances for efflorescence or deliquescence. Finally you will be asked to determine the amount of water lost by a sample of unknown hydrate on heating. From this amount, if given the formula or the molar mass of the anhydrous sample, you will be able to calculate the formula of the hydrate itself.

EXPERIMENTAL PROCEDURE   WEAR YOUR SAFETY GLASSES WHILE PERFORMING THIS EXPERIMENT

A. Identification of Hydrates.   Place about 0.5 g of each of the compounds listed below in small, dry test tubes, one compound to a tube. Observe carefully the behavior of each compound when you heat it gently with a burner flame. If droplets of water condense on the cool upper walls of the test tube, this is evidence that the compound may be a hydrate. Note the nature and color of the residue. Let the tube cool and try to dissolve the residue in a few cm^3 of water, warming very gently if necessary. A true hydrate will tend to dissolve in water, producing a solution with a color very similar to that of the original hydrate. If the compound is a carbohydrate, it will give off water on heating and will tend to char. The solution of the residue in water will often be caramel colored.

Nickel chloride   Sucrose
Potassium chloride   Potassium dichromate
Sodium tetraborate (borax)   Barium chloride

B. Reversibility of Hydration.   Gently heat a few crystals, ∼0.3 g, of hydrated cobalt(II) chloride, $CoCl_2 \cdot 6\ H_2O$, in an evaporating dish until the color change appears to be complete. Dissolve the residue in the evaporating dish in a few cm^3 of water from your wash bottle. Heat the resulting solution to boiling (**CAUTION!**), and carefully boil it to dryness. Note any color changes. Put the evaporating dish on the lab bench and let it cool.

C. Deliquescence and Efflorescence.   Place a few crystals of each of the compounds listed below on separate watch glasses and put them next to the dish of $CoCl_2$ prepared in Part B. Depending on their composition and the relative humidity (amount of moisture in the air), the samples may gradually lose water of hydration to, or pick up water from, the air. They may also remain unaffected. To establish whether the samples gain or lose mass, weigh each of them on a top-loading balance to 0.01 g. Record their masses. Weigh them again after about an hour to detect any change in mass. Observe the samples occasionally during the laboratory period, noting any changes in color, crystal structure, or degree of wetness that may occur.

$Na_2CO_3 \cdot 10\ H_2O$ (washing soda) $KAl(SO_4)_2 \cdot 12\ H_2O$ (alum)

$CaCl_2$ $CuSO_4$

D. Per Cent Water in a Hydrate.   Clean a porcelain crucible and its cover with 6 M HNO_3. Any stains that are not removed by this treatment will not interfere with this experiment. Rinse the crucible and cover with distilled water. Put the crucible with its cover slightly ajar on a clay triangle and heat with a burner flame, gently at first and then to redness for about 2 minutes. Allow the crucible and cover to cool, and then weigh them to 0.001 g on an analytical balance. Handle the crucible with clean crucible tongs.

Obtain a sample of unknown hydrate from the stockroom and place about a gram of sample in the crucible. Weigh the crucible, cover, and sample on the balance. Put the crucible on the clay triangle, with the cover in an off-center position to allow the escape of water vapor. Heat again, gently at first and then strongly, keeping the bottom of the crucible at red heat for about 10 minutes. Center the cover on the crucible and let it cool to room temperature. Weigh the cooled crucible along with its cover and contents.

Examine the solid residue. Add water until the crucible is two thirds full and stir. Warm gently if the residue does not dissolve readily. Does the residue appear to be soluble in water?

DATA AND OBSERVATIONS: Water of Hydration

A. Identification of Hydrates

	H$_2$O appears	Color of residue	Water soluble	Hydrate
Nickel chloride	_____	_____	_____	_____
Potassium chloride	_____	_____	_____	_____
Sodium tetraborate	_____	_____	_____	_____
Sucrose	_____	_____	_____	_____
Potassium dichromate	_____	_____	_____	_____
Barium chloride	_____	_____	_____	_____

B. Reversibility of Hydration

Summarize your observations on CoCl$_2 \cdot$ 6 H$_2$O:

Is the dehydration and hydration of CoCl$_2$ reversible?

C. Deliquescence and Efflorescence

	Mass (sample + glass) initial	final	Observations	Conclusions
Na$_2$CO$_3 \cdot$ 10 H$_2$O	_____	_____	_____	_____
KAl(SO$_4$)$_2 \cdot$ 12 H$_2$O	_____	_____	_____	_____
CaCl$_2$	_____	_____	_____	_____
CuSO$_4$	_____	_____	_____	_____
CoCl$_2$	_____	_____	_____	_____

Continued on following page   **39**

D.   Per Cent Water in a Hydrate

Mass of crucible and cover   _____ g

Mass of crucible, cover, and solid hydrate   _____ g

Mass of crucible, cover, and residue   _____ g

CALCULATIONS AND RESULTS

Mass of solid hydrate   _____ g

Mass of residue   _____ g

Mass of H_2O lost   _____ g

Percentage H_2O in the unknown hydrate   _____ %

Formula mass of anhydrous salt (if furnished)   _____

Number of moles of water per mole of unknown hydrate   _____

Unknown no.   _____

ADVANCE STUDY ASSIGNMENT: Water of Hydration

1. A student puts a sample of $Na_2SO_4 \cdot 10\,H_2O$ on a watch glass and observes it occasionally over a period of about an hour. She observes that the crystals gradually change from colorless, transparent, and reasonably large, to a fine white powder. She adds water to the powder and it dissolves. On evaporation of some of the water and subsequent cooling of the solution, large colorless crystals are produced.

    Explain these observations in light of the discussion section of this experiment.

2. The student is given a sample of a blue copper sulfate hydrate. She weighs the sample in a dry covered crucible and obtains a mass of 21.587 g for the crucible, cover, and sample. The mass of the empty crucible and cover had been found earlier to be 20.623 g. She then heats the crucible to drive off the water of hydration, keeping the crucible at red heat for 10 minutes with the cover slightly ajar. On cooling, she finds the mass of crucible, cover, and contents to be 21.240 g. The sample was converted in the process to very light blue anhydrous $CuSO_4$.

    a. What is the mass of the hydrate sample?

_____ g hydrate

    b. What is the mass of the anhydrous $CuSO_4$?

_____ g $CuSO_4$

    c. What is the mass of water driven off?

_____ g H_2O

    d. What is the per cent water in the hydrate?

$$\% \text{ water} = \frac{\text{mass of water in sample}}{\text{mass of hydrate sample}} \times 100\%$$

_____ %

    e. How many grams of water would there be in 100.0 g of hydrate? How many moles?

_____ grams H_2O _____ moles H_2O

Continued on following page   **41**

f. How many grams of $CuSO_4$ are there in 100.0 g of hydrate? How many moles? (What per cent of the hydrate is $CuSO_4$? Convert the mass of $CuSO_4$ to moles. Molar mass of $CuSO_4$ = 159.6 g.)

_____ grams $CuSO_4$ _____ moles $CuSO_4$

g. How many moles of water are present per mole of $CuSO_4$?

_____

h. What is the formula of the hydrate?

_____

EXPERIMENT

7 • Heat Effects and Calorimetry

Heat is a form of energy, sometimes called thermal energy, which can pass spontaneously from an object at a high temperature to an object at a lower temperature. If the two objects are in contact they will, given sufficient time, both reach the same temperature.

Heat flow is ordinarily measured in a device called a calorimeter. A calorimeter is simply a container with insulating walls, made so that essentially no heat is exchanged between the contents of the calorimeter and the surroundings. Within the calorimeter chemical reactions may occur or heat may pass from one part of the contents to another, but no heat flows into or out of the calorimeter from or to the surroundings.

A. Specific Heat. When heat flows into a substance the temperature of that substance will increase. The quantity of heat q required to cause a temperature change Δt of any substance is proportional to the mass m of the substance and the temperature change, as shown in Equation 1. The proportionality constant is called the specific heat of that substance.

$$q = (\text{specific heat}) \times m \times \Delta t = S.H. \times m \times \Delta t \qquad (1)$$

The specific heat can be considered to be the amount of heat required to raise the temperature of one gram of the substance by one degree Celsius (if you make m and Δt in Equation 1 both equal to one, then q will equal $S.H.$). Amounts of heat are measured in either joules or calories. To raise the temperature of one gram of water by one degree Celsius, 4.18 joules of heat must be added to the water. The specific heat of water is therefore 4.18 joules/g°C. Since 4.18 joules equals one calorie, we can also say that the specific heat of water is 1.00 calories/g°C. Ordinarily heat flow into or out of a substance is determined by the effect which that flow has on a known amount of water. Because water plays such an important role in these measurements the calorie, which was the unit of heat most commonly used until recently, was actually defined to be equal to the specific heat of water.

The specific heat of a metal can readily be measured in a calorimeter. A weighed amount of metal is heated to some known temperature and is then quickly poured into a calorimeter that contains a measured amount of water at a known temperature. Heat flows from the metal to the water, and the two equilibrate at some temperature between the initial temperatures of the metal and the water.

Assuming that no heat is lost from the calorimeter to the surroundings, and that a negligible amount of heat is absorbed by the calorimeter walls, the amount of heat that flows from the metal as it cools is equal to the amount of heat absorbed by the water.

In thermodynamic terms, the heat flow for the metal is equal in magnitude but opposite in direction, and hence in sign, to that for the water. For the heat flow q,

$$q_{H_2O} = -q_{metal} \qquad (2)$$

If we now express heat flow in terms of Equation 1 for both the water and the metal M, we get

$$S.H._{H_2O} m_{H_2O} \Delta t_{H_2O} = -S.H._M m_M \Delta t_M \qquad (3)$$

In this experiment we measure the masses of water and metal and their initial and final temperatures. (Note that $\Delta t_M < 0$ and $\Delta t_{H_2O} > 0$, since $\Delta t = t_{final} - t_{initial}$.) Given the specific heat of water we can find the positive specific heat of the metal by Equation 3. We will use this procedure to obtain the specific heat of an unknown metal.

The specific heat of a metal is related in a simple way to its atomic mass. Dulong and Petit discovered many years ago that about 25 joules were required to raise the temperature of one mole of many metals by one degree Celsius. This relation, shown in Equation 4, is known as the Law of Dulong and Petit:

$$AM \cong \frac{25}{S.H.(J/g°C)} \tag{4}$$

where *AM* is the atomic mass of the metal. Once the specific heat of the metal is known, the approximate atomic mass can be calculated by Equation 4. The Law of Dulong and Petit was one of the few rules available to early chemists in their studies of atomic masses.

B. Heat of Solution.   When a chemical reaction occurs in water solution, the situation is similar to that which is present when a hot metal sample is put into water. With such a reaction there is an exchange of heat between the reaction mixture and the solvent, water. As in the specific heat experiment, the heat flow for the reaction mixture is equal in magnitude but opposite in sign to that for the water. The heat flow associated with the reaction mixture is also equal to the enthalpy change, ΔH, for the reaction, so we obtain the equation

$$q_{reaction} = \Delta H_{reaction} = -q_{H_2O} \tag{5}$$

By measuring the mass of the water used as solvent, and by observing the temperature change that the water undergoes, we can find q_{H_2O} by Equation 1 and ΔH by Equation 5. If the temperature of the water goes up, heat has been *given off* by the reaction mixture, so the reaction is *exo*thermic; q_{H_2O} is positive and ΔH is *negative*. If the temperature of the water goes down, the reaction mixture has *absorbed heat from* the water and the reaction is *endo*thermic. In this case q_{H_2O} is negative and ΔH is *positive*. Both exo- and endothermic reactions are observed.

One of the simplest reactions that can be studied in solution occurs when a solid is dissolved in water. As an example of such a reaction note the solution of NaOH in water:

$$NaOH(s) \rightarrow Na^+(aq) + OH^-(aq); \Delta H = \Delta H_{solution} \tag{6}$$

When this reaction occurs, the temperature of the solution becomes much higher than that of the NaOH and water that were used. If we dissolve a known amount of NaOH in a measured amount of water in a calorimeter, and measure the temperature change that occurs, we can use Equation 1 to find q_{H_2O} for the reaction and use Equation 5 to obtain ΔH. Noting that ΔH is directly proportional to the amount of NaOH used, we can easily calculate $\Delta H_{solution}$ for either a gram or a mole of NaOH. In the second part of this experiment you will measure $\Delta H_{solution}$ for an unknown ionic solid.

EXPERIMENTAL PROCEDURE   WEAR YOUR SAFETY GLASSES WHILE PERFORMING THIS EXPERIMENT

A. Specific Heat.   From the stockroom obtain a calorimeter, a sensitive thermometer, a sample of metal in a large stoppered test tube, and a sample of unknown solid. (The thermometer is very expensive, so be careful when handling it.)

The calorimeter consists of two nested expanded polystyrene coffee cups fitted with a styrofoam cover. There are two holes in the cover for a thermometer and a glass stirring rod with a loop bent on one end. Assemble the experimental setup as shown in Figure 7.1.

Fill a 400-cm³ beaker two-thirds full of water and begin heating it to boiling. While the water is heating, weigh your sample of unknown metal in the large stoppered test tube to the nearest 0.1 g on a top loading or triple beam balance. Pour the metal into a dry container and weigh the empty test tube and stopper. Replace the metal in the test tube and put the *loosely* stoppered tube into the hot water in the beaker. The water level in the beaker should be high enough so that the top of the metal is below the water surface. Continue heating the metal in the water for at least 10

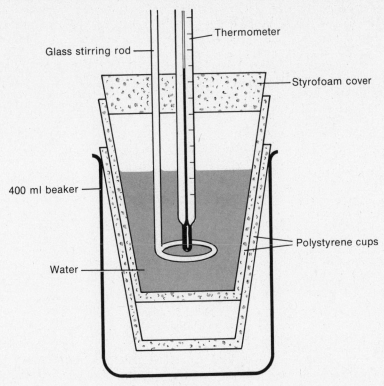

Glass stirring rod

Thermometer

Styrofoam cover

400 ml beaker

Polystyrene cups

Water

Figure 7.1

minutes after the water begins to boil to ensure that the metal attains the temperature of the boiling water. Add water as necessary to maintain the water level.

While the water is boiling, weigh the calorimeter to 0.1 g. Place about 40 cm³ of water in the calorimeter and weigh again. Insert the stirrer and thermometer into the cover and put it on the calorimeter. The thermometer bulb should be completely under the water.

Measure the temperature of the water in the calorimeter to 0.1°C. Take the test tube out of the beaker of boiling water, remove the stopper, and pour the metal into the water in the calorimeter. Be careful that no water adhering to the outside of the test tube runs into the calorimeter when you are pouring the metal. Replace the calorimeter cover and agitate the water as best you can with the glass stirrer. Record to 0.1°C the maximum temperature reached by the water. Repeat the experiment, using about 50 cm³ of water in the calorimeter. Be sure to dry your metal before reusing it; this can be done by heating the metal briefly in the test tube in boiling water and then pouring the metal onto a paper towel to drain. You can dry the hot test tube with a little compressed air.

The metal used in this part of the experiment is to be returned to the stockroom in the test tube in which you obtained it.

B. Heat of Solution. Place about 50 cm³ of distilled water in the calorimeter and weigh as in the previous procedure. Measure the temperature of the water to 0.1°C. The temperature should be within a degree or two of room temperature. In a small beaker weigh out about 5 g of the solid compound assigned to you. Make the weighing of the beaker and of the beaker plus solid to 0.1 g. Add the compound to the calorimeter. Stirring continuously and occasionally swirling the calorimeter, determine to 0.1°C the maximum or minimum temperature reached as the solid dissolves. Check to make sure that *all* the solid dissolved. A temperature change of at least five degrees should be obtained in this experiment. If necessary, repeat the experiment, increasing the amount of solid used.

Name _____ Section _____

DATA AND CALCULATIONS: Calorimetry

A. Specific Heat

	Trial 1	Trial 2
Mass of stoppered test tube plus metal	_____ g $\longrightarrow$	_____ g
Mass of test tube and stopper	_____ g $\longrightarrow$	_____ g
Mass of calorimeter	_____ g $\longrightarrow$	_____ g
Mass of calorimeter and water	_____ g	_____ g
Mass of water	_____ g	_____ g
Mass of metal	_____ g $\longrightarrow$	_____ g
Initial temperature of water in calorimeter	_____ °C	_____ °C
Initial temperature of metal (assume 100°C unless directed to do otherwise)	_____ °C $\longrightarrow$	_____ °C
Equilibrium temperature of metal and water in calorimeter	_____ °C	_____ °C
Δt_{water} $(t_{final} - t_{initial})$	_____ °C	_____ °C
Δt_{metal}	_____ °C	_____ °C
Specific heat of the metal (Eq. 3)	_____ J/g°C	_____ J/g°C
Approximate atomic mass of metal	_____	_____
Unknown no.	_____	

B. Heat of Solution

Mass of calorimeter plus water	_____ g
Mass of beaker	_____ g
Mass of beaker plus solid	_____ g

Continued on following page   **47**

Mass of water, m_{H_2O} _____ g

Mass of solid, m_S _____ g

Original temperature   _____ °C

Final temperature   _____ °C

q_{H_2O} for the reaction (Eq. 1) (S.H. = 4.18 J/g°C)

_____ joules

ΔH for the reaction (Eq. 5)

_____ joules

The quantity you have just calculated is approximately° equal to the heat of solution of your sample. Calculate the heat of solution per gram of solid sample.

$$\Delta H_{solution} = \text{_____} \text{ joules/g}$$

The solution reaction is endothermic exothermic. (Encircle correct answer.) Give your reasoning.

Solid unknown no. _____

(Optional)   Formula of compound used (if furnished by instructor)   _____

Mass of one mole of compound   _____ g

Heat of solution per mole of compound   _____ joules

° The value of ΔH will be approximate for several reasons. One of them is that we do not include the amount of heat absorbed by the solute. This effect is smaller than the likely experimental error, and thus we will ignore it.

ADVANCE STUDY ASSIGNMENT: Heat Effects and Calorimetry

1. A metal sample weighing 45.2 g and at a temperature of 100.0°C was placed in 38.6 g of water in a calorimeter at 25.2°C. At equilibrium the temperature of the water and metal was 33.0°C.

    a. What was Δt for the water? ($\Delta t = t_{final} - t_{initial}$)

_____ °C

    b. What was Δt for the metal?

_____ °C

    c. Taking the specific heat of water to be 4.18 J/g°C, calculate the specific heat of the metal, using Eq. 3.

_____ joules/g°C

    d. What is the approximate atomic mass of the metal? (Use Eq. 4.)

_____

2. When 2.0 g of NaOH were dissolved in 49.0 g water in a calorimeter at 24.0°C, the temperature of the solution went up to 34.5°C.

    a. Is this solution reaction exothermic? _____ Why?

    b. Calculate q_{H_2O}, using Eq. 1.

_____ joules

    c. Find ΔH for the reaction as it occurred in the calorimeter (Eq. 5).

$\Delta H =$ _____ joules

    d. Find ΔH for the solution of 1.00 g NaOH in water.

$\Delta H =$ _____ joules/g

    e. Find ΔH for the solution of one mole NaOH in water.

$\Delta H =$ _____ joules/mole

Continued on following page   **49**

f. Given that NaOH exists as Na$^+$ and OH$^-$ ions in solution, write the equation for the reaction that occurs when NaOH is dissolved in water.

g. Using enthalpies of formation as given in thermodynamic tables, calculate ΔH for the reaction in Part f and compare your answer with the result you obtained in Part e.

EXPERIMENT

8 • Molar Mass of a Volatile Liquid

One of the important applications of the Ideal Gas Law is found in the experimental determination of the molar masses of gases and vapors. In order to measure the molar mass of a gas or vapor we need simply to determine the mass of a given sample of the gas under known conditions of temperature and pressure. If the gas obeys the Ideal Gas Law,

$$PV = nRT \tag{1}$$

If the pressure P is in atmospheres, the volume V in liters, the temperature T in K, and the amount n in moles, then the gas constant R is equal to 0.0821 liter atm/(mole K).

The number of moles n is equal to the mass g of the gas divided by its molar mass in grams (MM). Substituting into Equation 1, we have

$$PV = \frac{gRT}{MM} \qquad\qquad MM = \frac{gRT}{PV} \tag{2}$$

This experiment involves measuring the molar mass of a volatile liquid by using Equation 2. A small amount of the liquid is introduced into a weighed flask. The flask is then placed in boiling water, where the liquid will vaporize completely, driving out the air and filling the flask with vapor at barometric pressure and the temperature of the boiling water. If we cool the flask so that the vapor condenses, we can measure the mass of the vapor and calculate a value for MM.

EXPERIMENTAL PROCEDURE*   WEAR YOUR SAFETY GLASSES WHILE PERFORMING THIS EXPERIMENT

Obtain a special round-bottomed flask, a stopper and cap, and an unknown liquid from the storeroom. Support the flask on an evaporating dish or in a beaker at all times. If you should break or crack the flask, report it to your instructor immediately so that it can be repaired. With the stopper loosely inserted in the neck of the flask, weigh the empty dry flask on the analytical balance. Use a copper loop, if necessary, to suspend the flask from the hook supporting the balance pan.

Pour about half your unknown liquid, about 5 ml, into the flask. Assemble the apparatus as shown in Figure 8.1. Place the cap on the neck of the flask. Add a few boiling chips to the water in the 600 ml beaker and heat the water to the boiling point. Watch the liquid level in your flask; the level should gradually drop as vapor escapes through the cap. After all the liquid has disappeared and no more vapor comes out of the cap, continue to boil the water gently for 5 to 8 minutes. Measure the temperature of the boiling water. Shut off the burner and wait until the water has stopped boiling (about ½ minute) and then loosen the clamp holding the flask in place. Slide out the flask, remove the cap, and *immediately* insert the stopper used previously.

Remove the flask from the beaker of water, holding it by the neck, which will be cooler. Immerse the flask in a beaker of cool water to a depth of about 5 cm. After holding the flask in the water for about two minutes to allow it to cool, carefully remove the stopper *for not more than a second or two* to allow air to enter, and again insert the stopper. (As the flask cools the vapor inside condenses and the pressure drops, which explains why air rushes in when the stopper is removed.)

---

* See W. L. Masterton and T. R. Williams, J. Chem. Educ. *36*, 528 (1959). For an alternate apparatus, see instructor's manual.

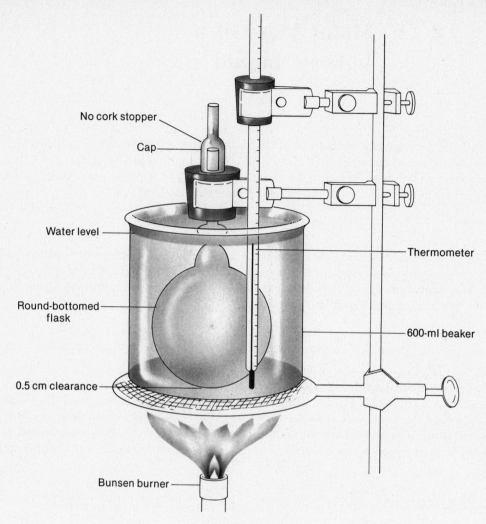

No cork stopper

Cap

Water level

Round-bottomed flask

0.5 cm clearance

Thermometer

600-ml beaker

Bunsen burner

Figure 8.1

Dry the flask with a towel to remove the surface water. Loosen the stopper momentarily to equalize any pressure differences, and reweigh the flask. Read the atmospheric pressure from the barometer.

Repeat the procedure using another 5 ml of your liquid sample.

You may obtain the volume of the flask from your instructor. Alternatively, he may direct you to measure its volume by weighing the flask stoppered and full of water on a top-loading balance. *Do not* fill the flask with water unless specifically told to do so.

When you have completed the experiment, return the flask to the storeroom; do not attempt to wash or clean it in any way.

DATA: Molar Mass of a Volatile Liquid

	Trial 1	Trial 2
Unknown no.	_____	
Mass of flask and stopper	_____ g	_____ g
Mass of flask, stopper, and condensed vapor	_____ g	_____ g
Mass of flask, stopper, and water (see directions)	_____ g	_____ g
Temperature of boiling water bath	_____ °C	_____ °C
Barometric pressure	_____ mm Hg	_____ mm Hg

CALCULATIONS AND RESULTS

	Trial 1	Trial 2
Pressure of vapor, P	_____ atm	_____ atm
Volume of flask (volume of vapor), V	_____ lit	_____ lit
Temperature of vapor, T	_____ K	_____ K
Mass of vapor, g	_____ g	_____ g
Molar mass of unknown, as found by substitution into Equation 2	_____ g	_____ g

ADVANCE STUDY ASSIGNMENT: Molar Mass of a Volatile Liquid

1. A student weighs an empty flask and stopper and finds the mass to be 54.868 g. She then adds about 5 ml of an unknown liquid and heats the flask in a boiling water bath at 99°C. After all the liquid is vaporized, she removes the flask from the bath, stoppers it, and lets it cool. After it is cool, she momentarily removes the stopper, then replaces it and weighs the flask and condensed vapor, obtaining a mass of 55.496 g. The volume of the flask is known to be 235.7 ml. The barometric pressure in the laboratory that day is 738 mm Hg.

    a. What was the pressure of the vapor in the flask in atm?

$$P = \text{_____} atm$$

    b. What was the temperature of the vapor in K? the volume of the flask in liters?

$$T = \text{_____} K \qquad V = \text{_____} l$$

    c. What was the mass of vapor that was present in the flask?

$$g = \text{_____} grams$$

    d. What is the mass of one mole of vapor? (Eq. 2.)

$$MM = \text{_____} grams/mole$$

2. How would each of the following procedural errors affect the results to be expected in this experiment? Give your reasoning in each case.

    a. All of the liquid was not vaporized when the flask was removed from the water bath.

    b. The flask was not dried before the final weighing with the condensed vapor inside.

    c. The flask was left open to the atmosphere while it was being cooled, and the stopper was inserted just before the final weighing.

    d. The flask was removed from the bath before the vapor had reached the temperature of the boiling water.

EXPERIMENT

9 • Analysis of an Aluminum-Zinc Alloy*

Some of the more active metals will react readily with solutions of strong acids, producing hydrogen gas and a solution of a salt of the metal. Small amounts of hydrogen are commonly prepared by the action of hydrochloric acid on metallic zinc:

$$Zn(s) + 2\,H^+(aq) \rightarrow H_2(g) + Zn^{2+}(aq) \tag{1}$$

From this equation it is clear that one mole of zinc produces one mole of hydrogen gas in this reaction. If the hydrogen were collected under known conditions, it would be possible to calculate the mass of zinc in a pure sample by measuring the amount of hydrogen it produced on reaction with acid.

Since aluminum reacts spontaneously with strong acids in a manner similar to that shown by zinc,

$$2\,Al(s) + 6\,H^+(aq) \rightarrow 2\,Al^{3+}(aq) + 3\,H_2(g) \tag{2}$$

we could find the amount of aluminum in a pure sample by measuring the amount of hydrogen produced by its reaction with an acid solution. In this case two moles of aluminum would produce three moles of hydrogen.

Since the amount of hydrogen produced by a gram of zinc is not the same as the amount produced by a gram of aluminum,

$$1 \text{ mole Zn} \rightarrow 1 \text{ mole } H_2, 65.4 \text{ g Zn} \rightarrow 1 \text{ mole } H_2, 1.00 \text{ g Zn} \rightarrow 0.0153 \text{ mole } H_2 \tag{3}$$

$$2 \text{ moles Al} \rightarrow 3 \text{ moles } H_2, 54.0 \text{ g Al} \rightarrow 3 \text{ moles } H_2, 1.00 \text{ g Al} \rightarrow 0.0556 \text{ mole } H_2 \tag{4}$$

it is possible to react an alloy of zinc and aluminum of known mass with acid, determine the amount of hydrogen gas evolved, and calculate the percentages of zinc and aluminum in the alloy, using Relations 3 and 4. The object of this experiment is to make such an analysis.

In this experiment you will react a weighed sample of an aluminum-zinc alloy with an excess of acid and collect the hydrogen gas evolved over water (Fig. 9.1). If you measure the volume, temperature, and total pressure of the gas and use the Ideal Gas Law, taking proper account of the pressure of water vapor in the system, you can calculate the number of moles of hydrogen produced by the sample:

$$P_{H_2}V = n_{H_2}RT, \quad n_{H_2} = \frac{P_{H_2}V}{RT} \tag{5}$$

The volume V and the temperature T of the hydrogen are easily obtained from the data. The pressure exerted by the dry hydrogen P_{H_2} requires more attention. The total pressure P of gas in the bottle is, by Dalton's Law, equal to the partial pressure of the hydrogen P_{H_2} plus the partial pressure of the water vapor P_{H_2O}:

$$P = P_{H_2} + P_{H_2O} \tag{6}$$

---

* W. L. Masterton, J. Chem. Educ. 38, 558 (1961). For a source of alloy samples, see instructor's manual.

The water vapor in the bottle is present with liquid water, so the gas is saturated with water vapor; the pressure P_{H_2O} under these conditions is equal to the vapor pressure VP_{H_2O} of water at the temperature of the experiment. This value is constant at a given temperature, and will be found in Appendix I at the end of this manual. The total gas pressure P in the flask is very nearly equal to the barometric pressure P_{bar}.[*]

Substituting these values into (6) and solving for P_{H_2}, we obtain

$$P_{H_2} = P_{bar} - VP_{H_2O} \tag{7}$$

Using (5), you can now calculate n_{H_2}, the number of moles of hydrogen produced by your weighed sample. You can then calculate the percentages of Al and Zn in the sample by properly applying (3) and (4) to your results. For a sample containing g_{Al} grams Al and g_{Zn} grams Zn, it follows that

$$n_{H_2} = (g_{Al} \times 0.0556) + (g_{Zn} \times 0.0153) \tag{8}$$

For a one gram sample, g_{Al} and g_{Zn} represent the mass fractions of Al and Zn, that is, % Al/100 and % Zn/100. Therefore

$$N_{H_2} = \left(\frac{\% \text{ Al}}{100} \times 0.0556\right) + \left(\frac{\% \text{ Zn}}{100} \times 0.0153\right) \tag{9}$$

where N_{H_2} = number of moles of H_2 produced *per gram* of sample.

Since it is also true that

$$\% \text{ Zn} = 100 - \% \text{ Al} \tag{10}$$

(9) can be written in the form

$$N_{H_2} = \left(\frac{\% \text{ Al}}{100} \times 0.0556\right) + \left(\frac{100 - \% \text{ Al}}{100} \times 0.0153\right) \tag{11}$$

We can solve Equation 11 directly for % Al if we know the number of moles of H_2 evolved per gram of sample. To save time in the laboratory and to avoid arithmetic errors, it is highly desirable to prepare in advance a graph giving N_{H_2} as a function of % Al. Then when N_{H_2} has been determined in the experiment, % Al in the sample can be read directly from the graph. Directions for preparing such a graph are given in Problem 1 in the Advance Study Assignment.

EXPERIMENTAL PROCEDURE
WEAR YOUR SAFETY GLASSES WHILE
PERFORMING THIS EXPERIMENT

Obtain a suction flask, large test tube, stopper assemblies and a sample of Al-Zn alloy from the stockroom. Assemble the apparatus as shown in Figure 9.1.

Take a gelatin capsule from the supply on the lab bench and weigh it on the analytical balance to ±0.0001 g. Pour your alloy sample out on a piece of paper and add about half of it to the capsule. If necessary, break up the turnings into smaller pieces by simply tearing them. Cover the capsule and weigh it again. The mass of sample should be between 0.1000 and 0.2000 grams. Use care in both weighings, since the sample is small and a small weighing error will produce a large experimental error. Put the remaining alloy back in its container.

Fill the suction flask and beaker about $\frac{2}{3}$ full of water. Moisten the stopper on the suction flask and insert it firmly into the flask. Open the pinch clamp and apply suction to the tubing attached

---

[*] In principle a small correction should be made for the difference in heights of the water levels inside and outside the suction flask. In practice the error made by neglecting this effect is much smaller than other experimental errors.

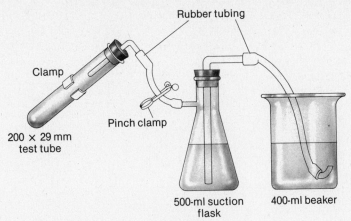

Figure 9.1

to the side arm of the suction flask. Pull water into the flask from the beaker until the water level in the flask is about 4 or 5 cm below the side arm. To apply suction, use a suction bulb or a short piece of rubber tubing attached temporarily to the tube that goes through the test tube stopper. Close the pinch clamp to prevent siphoning. The tubing from the beaker to the flask should be full of water, with no air bubbles.

Carefully remove the tubing from the beaker and put the end on the lab bench. As you do this, no water should leak out of the end of the tubing. Pour the water remaining in the beaker into another beaker, letting the 400-ml beaker drain for a second or two. Without drying it, weigh the empty beaker on a top-loading balance to ±0.1 g. Put the tubing back in this beaker.

Pour 10 ml of 6 M HCl, hydrochloric acid, as measured in your graduated cylinder, into the large test tube. Drop the gelatin capsule into the HCl solution; if it sticks to the tube, poke it down into the acid with your stirring rod. Insert the stopper firmly into the test tube and open the pinch clamp. If a little water goes into the beaker at that point, pour that water out, letting the beaker drain for a second or two.

Within 3 or 4 minutes the acid will eat through the wall of the capsule and begin to react with the alloy. The hydrogen gas that is formed will go into the suction flask and displace water from the flask into the beaker. The volume of water that is displaced will equal the volume of gas that is produced. As the reaction proceeds you will probably observe a dark foam, which contains particles of unreacted alloy. The foam may carry some of the alloy up the tube. Wiggle the tube gently to make sure that all of the alloy gets into the acid solution. The reaction should be over within five to ten minutes. At that time the liquid solution will again be clear, the foam will be essentially gone, the capsule will be all dissolved, and there should be no unreacted alloy. When the reaction is over, close the pinch clamp and take the tubing out of the beaker. Weigh the beaker and the displaced water to ±0.1 g. Measure the temperature of the water and the barometric pressure.

Pour the acid solution into the sink. Reassemble the apparatus and repeat the experiment with the remaining sample of alloy.

DATA: Analysis of an Aluminum-Zinc Alloy

	Trial 1	Trial 2
Mass of gelatin capsule	_____ g	_____ g
Mass of alloy sample plus capsule	_____ g	_____ g
Mass of empty beaker	_____ g	_____ g
Mass of beaker plus displaced water	_____ g	_____ g
Barometric pressure	_____ mm Hg	
Temperature	_____ °C	

CALCULATIONS

	Trial 1	Trial 2
Mass of alloy sample	_____ g	_____ g
Mass of displaced water	_____ g	_____ g
Volume of displaced water ($d = 1.00$ g/ml)	_____ ml	_____ ml
Volume of H_2, V	_____ liters	_____ liters
Temperature of H_2, T	_____ K	
Vapor pressure of water at T, VP_{H_2O}, from Appendix I	_____ mm Hg	
Pressure of dry H_2, P_{H_2}	_____ mm Hg;	_____ atm
Moles H_2 from sample, n_{H_2}	_____ moles	_____ moles
Moles H_2 per gram of sample, N_{H_2}	_____ moles/g	_____ moles/g
%Al (read from graph)	_____ %	_____ %
Unknown no.	_____	

ADVANCE STUDY ASSIGNMENT: Analysis of an Aluminum-Zinc Alloy

1.   On the following page, construct a graph of N_{H_2} vs. % Al. To do this, refer to Equation 11 and the discussion preceding it. Note that a plot of N_{H_2} vs. % Al should be a straight line (why?). To fix the position of a straight line it is necessary to locate only two points. The most obvious way to do this is to find N_{H_2} when % Al = 0 and when % Al = 100. If you wish you may calculate some intermediate points (for example, N_{H_2} when % Al = 50, or 20, or 70); all these points should lie on the same straight line.

2.   A student obtained the following data in this experiment. Fill in the blanks in the data and make the indicated calculations:

Mass of gelatin capsule	0.1168 g	Temperature, t	21°C
Mass of capsule plus alloy sample	0.2754 g	Temperature, T	_____ K
Mass of alloy sample, m	_____ g	Barometric pressure	746 mm Hg
Mass of empty beaker	141.2 g	Vapor pressure of H_2O at t (Appendix I)	_____ mm Hg
Mass of beaker plus displaced water	307.7 g		
Mass of displaced water	_____ g	Pressure of dry H_2, P_{H_2} (Eqn 7)	_____ mm Hg
Volume of displaced water (density = 1.00 g/ml)	_____ ml	Pressure of dry H_2	_____ atm

Volume, V, of H_2 = Volume of displaced water _____ ml; _____ liters

Find the number of moles of H_2 evolved, n_{H_2} (Eqn 5; V in liters, P_{H_2} in atm, T in K, $R = 0.0821$ liter-atm/mole K).

_____ moles H_2

Find N_{H_2}, the number of moles of H_2 per gram of sample (n_{H_2}/m). _____ moles H_2/g

Find the % Al in the sample from the graph prepared for Problem 1. _____ %Al

Find the % Al in the sample by using Equation 11. _____ %Al

Analysis of an Aluminum-Zinc Alloy
(Advance Study Assignment)

N_{H_2}

0.050

0.040

0.030

0.020

0.010

25%   50%   75%   100%

Per cent Al

EXPERIMENT

10 • The Atomic Spectrum of Hydrogen

When atoms are excited, either in an electric discharge or with heat, they tend to give off light. The light is emitted only at certain wavelengths which are characteristic of the atoms in the sample. These wavelengths constitute what is called the atomic spectrum of the excited element and reveal much of the detailed information we have regarding the electronic structure of atoms.

Atomic spectra are interpreted in terms of quantum theory. According to this theory, atoms can exist only in certain states, each of which has an associated fixed amount of energy. When an atom changes its state, it must absorb or emit an amount of energy which is just equal to the difference between the energies of the initial and final states. This energy may be absorbed or emitted in the form of light. The emission spectrum of an atom is obtained when excited atoms fall from higher to lower energy levels. Since there are many such levels, the atomic spectra of most elements are very complex.

Light is absorbed or emitted by atoms in the form of photons, each of which has a specific amount of energy, ϵ. This energy is related to the wavelength of light by the equation

$$\epsilon_{photon} = \frac{hc}{\lambda} \tag{1}$$

where h is Planck's constant, 6.62618×10^{-34} joule seconds, c is the speed of light, 2.997925×10^8 meters per second, and λ is the wavelength, in meters. The energy ϵ_{photon} is in joules and is the energy given off by one atom when it jumps from a higher to a lower energy level. Since total energy is conserved, the change in energy of the atom, $\Delta\epsilon_{atom}$, must equal the energy of the photon emitted:

$$\Delta\epsilon_{atom} = \epsilon_{photon} \tag{2}$$

where $\Delta\epsilon_{atom}$ is equal to the energy in the upper level minus the energy in the lower one. Combining Equations 1 and 2, we obtain the relation between the change in energy of the atom and the wavelength of light associated with that change:

$$\Delta\epsilon_{atom} = \epsilon_{upper} - \epsilon_{lower} = \epsilon_{photon} = \frac{hc}{\lambda} \tag{3}$$

The amount of energy in a photon given off when an atom makes a transition from one level to another is very small, of the order of 1×10^{-19} joules. This is not surprising since, after all, atoms are very small particles. To avoid such small numbers, we will work with one mole of atoms, much as we do in dealing with energies involved in chemical reactions. To do this we need only to multiply Equation 3 by Avogadro's number, N:

Let $$N\Delta\epsilon = \Delta E = N\epsilon_{upper} - N\epsilon_{lower} = E_{upper} - E_{lower} = \frac{Nhc}{\lambda}$$

Substituting the values for N, h, and c, and expressing the wavelength in nanometers rather than meters (1 meter = 1×10^9 nanometers), we obtain an equation relating energy change in kilojoules per mole of atoms to the wavelength of photons associated with such a change:

$$\Delta E = \frac{6.02205 \times 10^{23} \times 6.62618 \times 10^{-34} \text{ J sec} \times 2.997925 \times 10^8 \text{ m/sec}}{\lambda \text{(in nm)}} \times \frac{1 \times 10^9 \text{ nm}}{1 \text{ m}} \times \frac{1 \text{ kJ}}{1000 \text{ J}}$$

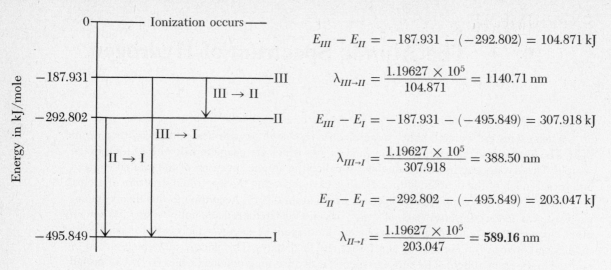

Figure 10.1.   Calculation of wavelengths of spectral lines from energy levels of the sodium atom.

$$\Delta E = E_{upper} - E_{lower} = \frac{1.19627 \times 10^5 \text{ kJ/mole}}{\lambda \text{ (in nm)}} \qquad \text{or} \qquad \lambda \text{ (in nm)} = \frac{1.19627 \times 10^5}{\Delta E \text{ (in kJ/mole)}} \qquad (4)$$

Equation 4 is useful in the interpretation of atomic spectra. Say, for example, we study the atomic spectrum of sodium and find that the wavelength of the strong yellow line is 589.16 nm (see Fig. 10.1). This line is known to result from a transition between two of the three lowest levels in the atom. The energies of these levels are shown in the figure. To make the determination of the levels which give rise to the 589.16 nm line, we note that there are three possible transitions, shown by downward arrows in the figure. We find the wavelengths associated with those transitions by first calculating ΔE ($E_{upper} - E_{lower}$) for each transition. Knowing ΔE we calculate λ by Equation 4. Clearly, the II $\rightarrow$ I transition is the source of the yellow line in the spectrum.

The simplest of all atomic spectra is that of the hydrogen atom. In 1886 Balmer showed that the lines in the spectrum of the hydrogen atom had wavelengths that could be expressed by a rather simple equation. Bohr, in 1913, explained the spectrum on a theoretical basis with his famous model of the hydrogen atom. According to Bohr's theory, the energies allowed to a hydrogen atom are given by the equation

$$\epsilon_n = \frac{-B}{n^2} \qquad (5)$$

where B is a constant predicted by the theory and n is an integer, 1, 2, 3, . . . , called a quantum number. It has been found that all the lines in the atomic spectrum of hydrogen can be associated with energy levels in the atom which are predicted with great accuracy by Bohr's equation. When we write Equation 5 in terms of a mole of H atoms, and substitute the numerical value for B, we obtain

$$E_n = \frac{-1312.04}{n^2} \text{ kilojoules per mole, } n = 1, 2, 3, \ldots \qquad (6)$$

Using Equation 6 you can calculate, very accurately indeed, the energy levels for hydrogen. Transitions between these levels give rise to the wavelengths in the atomic spectrum of hydrogen. These wavelengths are also known very accurately. Given both the energy levels and the wavelengths, it is possible to determine the actual levels associated with each wavelength. In this experiment your task will be to make determinations of this type for the observed wavelengths in the hydrogen atomic spectrum which are listed in Table 10.1.

TABLE 10.1 SOME WAVELENGTHS (IN nm) IN THE SPECTRUM OF THE HYDROGEN ATOM AS MEASURED IN A VACUUM

Wavelength	Assignment $n_{hi} \longrightarrow n_{lo}$	Wavelength	Assignment $n_{hi} \longrightarrow n_{lo}$	Wavelength	Assignment $n_{hi} \longrightarrow n_{lo}$
97.25	_____	410.29	_____	1005.2	_____
102.57	_____	434.17	_____	1094.1	_____
121.57	_____	486.27	_____	1282.2	_____
389.02	_____	656.47	_____	1875.6	_____
397.12	_____	954.86	_____	4052.3	_____

EXPERIMENTAL PROCEDURE

There are several ways we might analyze an atomic spectrum, given the energy levels of the atom involved. A simple and effective method is to calculate the wavelengths of some of the lines arising from transitions between some of the lower energy levels, and see if they match those that are observed. We shall use this method in our experiment. All the data are good to at least five significant figures, so by using electronic calculators you should be able to make very accurate determinations.

A. Calculations of the Energy Levels of the Hydrogen Atom. Given the expression for E_n in Equation 6, it is possible to calculate the energy for each of the allowed levels of the H atom starting with $n = 1$. Using your calculator, calculate the energy in kJ/mole of each of the ten lowest levels of the H atom. Note that the energies are all negative, so that the *lowest* energy will have the *largest* allowed negative value. Enter these values in the table of energy levels, Table 10.2. On the energy level diagram provided, plot along the y axis each of the six lowest energies, drawing a horizontal line at the allowed level and writing the value of the energy alongside the line near the y axis. Write the quantum number associated with the level to the right of the line.

B. Calculation of the Wavelengths of the Lines in the Hydrogen Spectrum. The lines in the hydrogen spectrum all arise from jumps made by the atom from one energy level to another. The wavelengths in nm of these lines can be calculated by Equation 4, where ΔE is the difference in energy in kJ/mole between any two allowed levels. For example, to find the wavelength of the spectral line associated with a transition from the $n = 2$ level to the $n = 1$ level, calculate the difference, ΔE, between the energies of those two levels. Then substitute ΔE into Equation 4 to obtain this wavelength in nanometers.

Using the procedure we have outlined, calculate the wavelengths in nm of all the lines we have indicated in Table 10.3. That is, calculate the wavelengths of all the lines that can arise from transitions between any two of the six lowest levels of the H atom. Enter these values in Table 10.3.

C. Assignment of Observed Lines in the Hydrogen Spectrum. Compare the wavelengths you have calculated with those which are listed in Table 10.1. If you have made your calculations properly, your wavelengths should match, within the error of your calculation, several of those which are observed. On the line opposite each wavelength in Table 10.1, write the quantum numbers of the upper and lower states for each line whose origin you can recognize by comparison of your calculated values with the observed values. On the energy level diagram, draw a vertical arrow pointing down (light is emitted, $\Delta E < 0$) between those pairs of levels which you associate with any of the observed wavelengths. By each arrow write the wavelength of the line originating from that transition.

There are a few wavelengths in Table 10.1 which have not yet been calculated. By assignments already made and by an examination of the transitions you have marked on the diagram, deduce the quantum states that are likely to be associated with the as yet unassigned lines. This is perhaps most easily done by first calculating the value of ΔE, which is associated with a given wavelength. Then find two values of E_n whose difference is equal to ΔE. The quantum numbers for the two E_n states whose energy difference is ΔE will be the ones which are to be assigned to the given wavelength. When you have found n_{hi} and n_{lo} for a wavelength, write them in Table 10.1; continue until all the lines in the table have been assigned.

D.   The Balmer Series.   This is the most famous series in the atomic spectrum of hydrogen. Carry out calculations in connection with this series as directed in the Data and Calculations section.

DATA AND CALCULATIONS: The Atomic Spectrum of Hydrogen

A. The Energy Levels of the Hydrogen Atom

Energies are to be calculated from Equation 6 for the ten lowest energy states.

TABLE 10.2

Quantum Number, n	Energy, E_n, in kJ/mole	Quantum Number, n	Energy, E_n, in kJ/mole
___	_____	___	_____
___	_____	___	_____
___	_____	___	_____
___	_____	___	_____
___	_____	___	_____

B. Calculation of Wavelengths in the Spectrum of the H Atom

In the upper half of each box write ΔE, the difference in energy in kJ/mole between $E_{n_{hi}}$ and $E_{n_{lo}}$. In the lower half of the box, write λ in nm associated with that value of ΔE.

TABLE 10.3

$$\Delta E = E_{n_{hi}} - E_{n_{lo}}$$

$$\lambda \text{ (nm)} = \frac{1.19627 \times 10^5}{\Delta E}$$

Continued on following page

C. Assignment of Wavelengths

1. As directed in the procedure, assign n_{hi} and n_{lo} for each wavelength in Table 10.1 which corresponds to a wavelength calculated in Table 10.3.
2. List below any wavelengths you cannot yet assign and find their origin.

Wavelength λ observed	ΔE transition	Probable transition $n_{hi} \longrightarrow n_{lo}$	λ calculated in nm (Eq. 4)
_____	_____	_____	_____
_____	_____	_____	_____
_____	_____	_____	_____
_____	_____	_____	_____

D. 1. THE BALMER SERIES. When Balmer found his famous series for hydrogen in 1886, he was limited experimentally to wavelengths in the visible and near ultraviolet regions from 250 nm to 700 nm, so all the lines in his series lie in that region. On the basis of the entries in Table 10.3 and the transitions on your energy level diagram, what common characteristic do the lines in the Balmer Series have?

What would be the longest possible wavelength for a line in the Balmer series?

$$\lambda = \text{_____} nm$$

What would be the shortest possible wavelength that a line in the Balmer series could have? Hint: What is the largest possible value of ΔE to be associated with a line in the Balmer series?

$$\lambda = \text{_____} nm$$

Fundamentally, why would any line in the hydrogen spectrum between 250 mm and 700 nm belong to the Balmer series? Hint: On the energy level diagram note the range of possible values of ΔE for transitions to the $n = 1$ level and to the $n = 3$ level. Could a spectral line involving a transition to the $n = 1$ level have a wavelength in the range indicated?

2. THE IONIZATION ENERGY OF HYDROGEN. In the normal hydrogen atom the electron is in its lowest energy state, which is called the ground state of the atom. The maximum electronic energy that a hydrogen atom can have is 0 kJ/mole, at which point the electron would essentially be removed from the atom and it would become a H^+ ion. How much energy in kilojoules per mole does it take to ionize an H atom?

$$\text{_____} kJ/mole$$

Continued on following page

The ionization energy of hydrogen is often expressed in units other than kJ/mole. What would it be in joules per atom?

_____J/atom

(The energy level diagram to be completed in Part A is on the following page.)

The Atomic Spectrum of Hydrogen
Energy Level Diagram

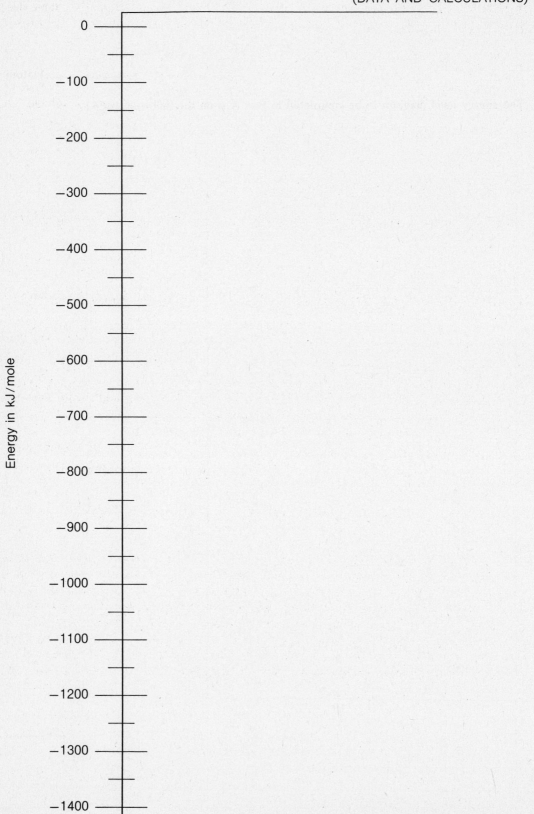

Energy in kJ/mole

0

−100

−200

−300

−400

−500

−600

−700

−800

−900

−1000

−1100

−1200

−1300

−1400

ADVANCE STUDY ASSIGNMENT: The Atomic Spectrum of Hydrogen

1.   The helium ion, He^+, has energy levels which are similar to those of the hydrogen atom, since both species have only one electron. The energy levels of the He^+ ion are given by the equation

$$E_n = -\frac{5248.16}{n^2} \text{ kJ/mole} \qquad n = 1, 2, 3, \ldots$$

a. Calculate the energies in kJ/mole for the four lowest energy levels of the He^+ ion.

$E_1 = $ _____ kJ/mole

$E_2 = $ _____ kJ/mole

$E_3 = $ _____ kJ/mole

$E_4 = $ _____ kJ/mole

b. One of the most important transitions for the He^+ ion involves a jump from the $n = 2$ to the $n = 1$ level. ΔE for this transition equals $E_2 - E_1$, where these two energies are obtained as in Part a. Find the value of ΔE in kJ/mole. Find the wavelength in nm of the line emitted when this transition occurs; use Equation 4 to make the calculation.

$\Delta E = $ _____ kJ/mole; $\lambda = $ _____ nm

c. Three of the strongest lines in the He^+ ion spectrum are observed at the following wavelengths: (1) 121.57 nm; (2) 164.12 nm; (3) 468.90 nm. Find the quantum numbers of the initial and final states for the transitions which give rise to these three lines. Do this by calculating, using Equation 4, the wavelengths of lines which can originate from transitions involving any two of the four lowest levels. You calculated one such wavelength in Part b. Make similar calculations with the other possible pairs of levels. When a calculated wavelength matches an observed one, write down n_{hi} and n_{lo} for that line. Continue until you have assigned all three of the lines. Make your calculations on the other side of this page.

(1) _____→_____   (2) _____→_____   (3) _____→_____

75

EXPERIMENT

11 • The Alkaline Earths and the Halogens—Two Families in the Periodic Table

The Periodic Table arranges the elements in order of increasing atomic number in horizontal rows of such length that elements with similar properties recur periodically; that is, they fall directly beneath each other in the Table. The elements in a given vertical column are referred to as a family or group. The physical and chemical properties of the elements in a given family change gradually as one goes from one element in the column to the next. By observing the trends in properties the elements can be arranged in the order in which they appear in the Periodic Table. In this experiment we will study the properties of the elements in two families in the Periodic Table, the alkaline earths (Group 2) and the halogens (Group 7).

The alkaline earths are all moderately reactive metals and include barium, beryllium, calcium, magnesium, radium, and strontium. (Since beryllium compounds are rarely encountered and often very poisonous, and radium compounds are highly radioactive, we will not include these two elements in this experiment.) All the alkaline earths exist in their compounds and in solution as M^{2+} cations (Mg^{2+}, Ca^{2+}, etc.). If solutions of these cations are mixed with solutions containing X^{2-} anions (CO_3^{2-}, SO_4^{2-}, etc.), salts of the alkaline earths will precipitate if the compound MX is insoluble:

$$M^{2+}(aq) + X^{2-}(aq) \rightarrow MX(s) \qquad \text{if MX is insoluble} \qquad (1)$$

$$M^{2+} = Ba^{2+}, Ca^{2+}, Mg^{2+}, \text{ or } Sr^{2+}; X^{2-} = SO_4^{2-}, CO_3^{2-}, C_2O_4^{2-}, \text{ or } CrO_4^{2-}.$$

We would expect, and indeed observe, that the solubilities of the salts of the alkaline earth cations with any one of the given anions show a smooth trend consistent with the order of the cations in the Periodic Table. That is, as we go from one end of the alkaline earth family to the other, the solubilities of, say, the sulfate salts either gradually increase or decrease. Similar trends exist for the carbonates, oxalates, and chromates formed by those cations. By determining such trends in this experiment, you will be able to confirm the order of the alkaline earths in the Periodic Table.

The elementary halogens are also relatively reactive. They include astatine, bromine, chlorine, fluorine, and iodine. We will not study astatine and fluorine in this experiment, since the former is radioactive and the latter is too reactive to be safe. Unlike the alkaline earths, the halogen atoms tend to gain electrons, forming X^- anions (Cl^-, Br^-, etc.). Because of this property, the halogens are oxidizing agents, species which tend to oxidize (remove electrons from) other species. An interesting and simple example of the sort of reaction that may occur arises when a solution containing a halogen (Cl_2, Br_2, I_2) is mixed with a solution containing a halide ion (Cl^-, Br^-, I^-). Taking X_2 to be the halogen, and Y^- to be a halide ion, the following reaction may occur:

$$X_2(aq) + 2\ Y^-(aq) \rightarrow 2\ X^-(aq) + Y_2(aq) \qquad (2)$$

The reaction will occur if X_2 is a better oxidizing agent than Y_2, since then X_2 can produce Y_2 by removing electrons from the Y^- ions. If Y_2 is a better oxidizing agent than X_2, Reaction 2 will not proceed but will be spontaneous in the opposite direction.

In this experiment we will mix solutions of halogens and halide ions to determine the relative oxidizing strengths of the halogens. These strengths show a smooth variation as one goes from one halogen to the next in the Periodic Table. We will be able to tell if a reaction occurs by the colors we observe. In water, and particularly in some organic solvents, the halogens have characteristic colors. The halide ions are colorless in water solution and insoluble in organic solvents. Bromine

77

(Br_2) in hexane, C_6H_{14}(HEX), is orange, while Cl_2 and I_2 in that solvent have quite different colors.

Say, for example, we shake a water solution of Br_2 with a little hexane, which is lighter than and insoluble in water. The Br_2 is much more soluble in HEX than in water and goes into the HEX layer, giving it an orange color. To that mixture we add a solution containing a halide ion, say Cl^- ion, and mix well. If Br_2 is a better oxidizing agent than Cl_2, it will take electrons from the chloride ions and will be converted to bromide, Br^-, ions; the reaction would be

$$Br_2(aq) + 2\ Cl^-(aq) \rightarrow 2\ Br^-(aq) + Cl_2(aq) \tag{3}$$

If the reaction occurs, the color of the HEX layer will of necessity change, since Br_2 will be used up and Cl_2 will form. The color of the HEX layer will go from orange to that of a solution of Cl_2 in HEX. If the reaction does *not* occur, the color of the HEX layer will remain orange. By using this line of reasoning, and by working with the possible mixtures of halogens and halide ions, you should be able to arrange the halogens in order of increasing oxidizing power, which must correspond to their order in the Periodic Table.

One difficulty that you may have in this experiment involves terminology rather than actual chemistry. You must learn to distinguish the halogen *elements* from the halide *ions*, since the two kinds of species are not at all the same, even though their names are similar:

Elementary halogens	Halide Ions
Bromine, Br_2	Bromide ion, Br^-
Chlorine, Cl_2	Chloride ion, Cl^-
Iodine, I_2	Iodide ion, I^-

The *halogens* are molecular substances and oxidizing agents, and all have odors. They are only slightly soluble in water and are much more soluble in HEX, where they have distinct colors. The *halide ions* exist in solution only in water, have no color or odor, and are *not* oxidizing agents. They do not dissolve in HEX.

Given the solubility properties of the alkaline earth cations, and the oxidizing power of the halogens, it is possible to develop a systematic procedure for determining the presence of any Group 2 cation and any Group 7 anion in a solution. In the last part of this experiment you will be asked to set up such a procedure and use it to establish the identity of an unknown solution containing a single alkaline earth halide.

EXPERIMENTAL PROCEDURE
WEAR YOUR SAFETY GLASSES WHILE PERFORMING THIS EXPERIMENT

I. Relative Solubilities of Some Salts of the Alkaline Earths. To each of four small test tubes add about 1 ml (approximately 12 drops) of 1 M H_2SO_4. Then add 1 ml of 0.1 M solutions of the nitrate salts of barium, calcium, magnesium, and strontium to those tubes, one solution to a tube. Stir each mixture with your glass stirring rod, rinsing the rod in a beaker of distilled water between stirs. Record your results on the solubilities of the sulfates of the alkaline earths in the Table, noting whether a precipitate forms, and any characteristics (such as color, amount, size of particles and settling tendencies) that might distinguish it.

Rinse out the test tubes, and to each add 1 ml 1 M Na_2CO_3. Then add 1 ml of the solutions of the alkaline earth salts, one solution to a tube, as before. Record your observations on the solubility properties of the carbonates of the alkaline earth cations. Rinse out the tubes, and test for the solubilities of the oxalates of these cations, using 0.25 M $(NH_4)_2C_2O_4$ as the precipitating reagent. Finally, determine the relative solubilities of the chromates of the alkaline earths, using 1 ml 1 M K_2CrO_4 plus 1 ml 1 M acetic acid as the test reagent.

II. Relative Oxidizing Powers of the Halogens. In a small test tube place a few ml of bromine-saturated water and add 1 ml of hexane. Stopper the test tube and shake until the bromine color is mostly in the HEX layer. (*CAUTIONS:* Avoid breathing the halogen vapors. Don't use your finger to stopper the tube, since a halogen solution can give you a bad chemical burn.) Repeat the experiment using chlorine water and iodine water with separate samples of HEX, noting any color changes as the bromine, chlorine, and iodine are extracted from the water layer into the HEX layer.

To each of three small test tubes add 1 ml bromine water and 1 ml HEX. Then add 1 ml 0.1 M NaCl to the first tube, 1 ml 0.1 M NaBr to the second, and 1 ml 0.1 M NaI to the third. Stopper each tube and shake it. Note the color of the HEX phase above each solution. If the color is that of Br_2 in HEX, then no reaction between Br_2 and the added halide ion occurred. If the color is not that of Br_2 in HEX, a reaction indeed occurred, and Br_2 oxidized that anion, producing the halogen. In such a case, Br_2 is a stronger oxidizing agent than the halogen that was produced.

Rinse out the tubes, and this time add 1 ml chlorine water and 1 ml HEX to each tube. Then add 1 ml of the 0.1 M solutions of the sodium halide salts, one solution to a tube, as before. Stopper each tube and shake, noting the color of the HEX layer after shaking. Depending on whether the color is that of Cl_2 in HEX or not, decide whether Cl_2 is a better oxidizing agent than Br_2 or I_2. Again, rinse out the tubes, and add 1 ml iodine water and 1 ml HEX to each. Test each tube with 1 ml of a sodium halide salt solution, and determine whether I_2 is able to oxidize Cl^- or Br^- ions. Record all your observations in the Table in Part II.

III. Identification of an Alkaline Earth Halide. Your observations on the solubility properties of the alkaline earth cations should allow you to develop a method for determining which of those cations is present in a solution containing one Group 2 cation and no other cations. The method will involve testing samples of the solution with one or more of the reagents you used in Part I. Indicate on the data page how you would proceed.

In a similar way you can determine which halide ion is present in a solution containing only one such anion and no others. There you will need to test a solution of an oxidizing halogen with your unknown to see how the halide ion is affected. From the behavior of the halogen-halide ion mixtures you studied in Part II you should be able to identify easily the particular halide that is present. Describe your method on the data page, obtain an unknown solution of an alkaline earth halide, and then use your procedure to determine the cation and anion that it contains.

DATA AND OBSERVATIONS: The Alkaline Earths and the Halogens

I. Solubilities of Salts of the Alkaline Earths

	1 M H_2SO_4	1 M Na_2CO_3	0.25 M $(NH_4)_2C_2O_4$	1 M K_2CrO_4 1 M Acetic Acid
$Ba(NO_3)_2$				
$Ca(NO_3)_2$				
$Mg(NO_3)_2$				
$Sr(NO_3)_2$				

Key: P = precipitate forms; S = no precipitate.

Note any distinguishing characteristics of precipitate.

Consider the relative solubilities of the Group 2 cations in the various precipitating reagents. On the basis of the trends you observed, list the four alkaline earths in the order in which they should appear in the Periodic Table. *Start with the one which forms the most soluble oxalate.*

_____ _____ _____ _____

Why did you arrange the elements as you did? Is the order consistent with the properties of the cations in all of the precipitating reagents?

II. Relative Oxidizing Powers of the Halogens

a. Color of the halogen in solution:

	Br_2	Cl_2	I_2
Water	_____	_____	_____
HEX	_____	_____	_____

Continued on following page

b. Reactions between halogens and halides:

	Br$^-$	Cl$^-$	I$^-$
Br$_2$			
Cl$_2$			
I$_2$			

State initial and final colors of HEX layer. R = reaction occurs; NR = no reaction occurs.

Rank the halogens in order of their increasing oxidizing power.

_____ _____ _____

Is this their order in the Periodic Table?

III. Identification of an alkaline earth halide

Procedure for identifying the Group 2 cation:

Procedure for identifying the Group 7 anion:

Observations on unknown alkaline earth halide solution:

Cation present _____

Anion present _____

Unknown no. _____

ADVANCE STUDY ASSIGNMENT: The Alkaline Earths and the Halogens

1. Carbon has a density of about 2.3 g/cm^3 when in the form of graphite. Germanium has a density of about 5.3 g/cm^3. Using the Periodic Table, predict whether silicon would have a density greater than that of germanium. Make the same kind of prediction for tin. Indicate your reasoning.

2. Substances A, B, and C can all act as oxidizing agents. In the reactions in which they participate, they are reduced to A$^-$, B$^-$, and C$^-$ ions. When a solution of A is mixed with one containing C$^-$ ions, an oxidation-reduction reaction occurs. Write the equation for the reaction:

Which species is oxidized? _____

Which is reduced? _____

When a solution of A is mixed with one containing B$^-$ ions, no reaction occurs.

Is A a better oxidizing agent than B? _____

Is A a better oxidizing agent than C? _____

Arrange A, B, and C in order of increasing strengths as oxidizing agents.

3. You are given an unknown solution which may contain only one salt from the following set: NaA, NaB, NaC. In solution each salt dissociates completely into Na$^+$ ion and the anion A$^-$, B$^-$, or C$^-$, whose properties are those in Problem 2. The Na$^+$ ion is effectively inert. Given a way to determine whether the reactions in Problem 2 occur, develop a simple procedure for identifying the salt present in your unknown.

EXPERIMENT

12 • The Geometrical Structure of Molecules—An Experiment Using Molecular Models

Many years ago it was observed that in many of its compounds the carbon atom formed four chemical linkages to other atoms. As early as 1870, graphic formulas of carbon compounds were drawn as shown:

$$
\begin{array}{cc}
\begin{array}{c}
\text{H} \\
| \\
\text{H}-\text{C}-\text{H} \\
| \\
\text{H}
\end{array}
&
\begin{array}{c}
\text{H} \quad \text{H} \\
| \quad\ | \\
\text{C}=\text{C} \\
| \quad\ | \\
\text{H} \quad \text{H}
\end{array}
\\
\text{methane} & \text{ethylene}
\end{array}
$$

Although such drawings as these would imply that the atom-atom linkages, indicated by valence strokes, lie in a plane, chemical evidence, particularly the existence of only one substance with the graphic formula

$$
\begin{array}{c}
\text{Cl} \\
| \\
\text{H}-\text{C}-\text{Cl} \\
| \\
\text{H}
\end{array}
$$

requires that the linkages be directed toward the corners of a tetrahedron, at the center of which is the carbon atom.

The physical significance of the chemical linkages between atoms, expressed by the lines or valence strokes in molecular structure diagrams, became evident soon after the discovery of the electron. In 1916 in a classic paper, G. N. Lewis suggested, on the basis of chemical evidence, that the single bonds in graphic formulas involve two electrons and that an atom tends to hold eight electrons in its outermost or valence shell.

Lewis' proposal that atoms generally have eight electrons in their outer shell proved to be extremely useful and has come to be known as the octet rule. It can be applied to many atoms, but is particularly important in the treatment of covalent compounds of atoms in the second row of the Periodic Table. For atoms such as carbon, oxygen, nitrogen, and fluorine, the eight valence electrons occur in pairs that occupy tetrahedral positions around the central atom core. Some of the electron pairs do not participate directly in chemical bonding and are called unshared or nonbonding pairs; however, the structures of compounds containing such unshared pairs reflect the tetrahedral arrangement of the four pairs of valence shell electrons. In the H_2O molecule, which obeys the octet rule, the four pairs of electrons around the central oxygen atom occupy essentially tetrahedral positions; there are two unshared nonbonding pairs and two bonding pairs which are shared by the O atom and the two H atoms. The H—O—H bond angle is nearly but not exactly tetrahedral since the properties of shared and unshared pairs of electrons are not exactly alike.

$$
\begin{array}{c}
\cdot\ddot{\text{O}}\cdot \\
\diagup \quad \diagdown \\
\text{H} \qquad \text{H}
\end{array}
$$

Most molecules obey the octet rule. Essentially, all organic molecules obey the rule, and so do most inorganic molecules and ions. For species that obey the octet rule it is possible to draw electron-dot, or Lewis, structures. The drawing of the H_2O molecule just above is an example of a Lewis structure. Below we have listed several others:

$$
\begin{array}{cccccc}
& :\!\ddot{C}l\!: & & & & \\
& | & & & & \\
H\!-\!\!\overset{\displaystyle |}{\underset{\displaystyle |}{C}}\!\!-\!H & & H\!-\!\ddot{N}\!-\!H & :\ddot{O}\!-\!H^- & H\!-\!\overset{H}{\underset{H}{C}}\!-\!\overset{H}{\underset{H}{C}}\!-\!H & H\!-\!\overset{H}{C}\!=\!\overset{H}{C}\!-\!H \\
& :\!\ddot{C}l\!: & \quad H & & H\ H & H\ H
\end{array}
$$

In each of the above structures there are eight electrons around each atom (except for H atoms, which always have two electrons). There are two electrons in each bond. When counting electrons in these structures, one considers the electrons in a bond between two atoms as belonging to the atom under consideration. In the CH_2Cl_2 molecule just above, for example, the Cl atoms each have eight electrons, including the two in the single bond to the C atom. The C atom also has eight electrons, two from each of the four bonds to that atom. The bonding and non-bonding electrons in Lewis structures are all from the *outermost* shells of the atoms involved, and are the so-called valence electrons of those atoms. For the main group elements, the number of valence electrons in an atom is equal to the group number of the element in the Periodic Table. Carbon, in Group 4, has four valence electrons in its atoms; hydrogen, in Group 1, has one; chlorine, in Group 7, has seven valence electrons. In an octet rule structure the valence electrons from all the atoms are arranged in such a way that each atom, except hydrogen, has eight electrons.

Often it is quite easy to construct an octet rule structure for a molecule. Given that an oxygen atom has six valence electrons (Group 6) and a hydrogen atom has one, it is clear that one O and two H atoms have a total of eight valence electrons; the octet rule structure for H_2O, which we discussed earlier, follows by inspection. Structures like that of H_2O, involving only single bonds and non-bonding electron pairs are common. Sometimes, however, there is a "shortage" of electrons; that is, it is not possible to construct an octet rule structure in which all the electron pairs are either in single bonds or are nonbonding. C_2H_4 is a typical example of such a species. In such cases, octet rule structures can often be made in which two atoms are bonded by two pairs, rather than one pair, of electrons. The two pairs of electrons form a double bond. In the C_2H_4 molecule, shown above, the C atoms each get four of their electrons from the double bond. The assumption that electrons behave this way is supported by the fact that the C=C double bond is both shorter and stronger than the C—C single bond in the C_2H_6 molecule (see above). Double bonds, and triple bonds, occur in many molecules, usually between C, O, N, and/or S atoms.

Lewis structures can be used to predict molecular and ionic geometries. All that is needed is to assume that the four pairs of electrons around each atom are arranged tetrahedrally. We have seen how that assumption leads to the correct geometry for H_2O. Applying the same principle to the species whose Lewis structures we listed earlier, we would predict, correctly, that the $C_2H_2Cl_2$ molecule would be tetrahedral (roughly anyway), that NH_3 would be pyramidal (with the non-bonding electron pair sticking up from the pyramid made from the atoms, that the bond angles in C_2H_6 are all tetrahedral, and that the C_2H_4 molecule is planar (the two bonding pairs in the double bond are in a sort of banana bonding arrangement above and below the plane of the molecule. In describing molecular geometry we indicate the positions of the atomic nuclei, not the electrons. The NH_3 molecule is pyramidal, not tetrahedral.

It is also possible to predict polarity from Lewis structures. Polar molecules have their center of positive charge at a different point than their center of negative charge. This separation of charges produces a dipole moment in the molecule. Covalent bonds between different kinds of atoms are polar; all heteronuclear diatomic molecules are polar. In some molecules the polarity from one bond may be cancelled by that from others. Carbon dioxide, CO_2, which is linear, is a nonpolar molecule. Methane, CH_4, which is tetrahedral, is also nonpolar. Among the species whose Lewis structures we have listed, we find that H_2O, CH_2Cl_2, NH_3, and OH^- are polar. C_2H_6 and C_2H_4 are nonpolar.

For some molecules with a given molecular formula, it is possible to satisfy the octet rule with different atomic arrangements. A simple example would be

$$H-\overset{\overset{\displaystyle H}{|}}{\underset{\underset{\displaystyle H}{|}}{C}}-\overset{\overset{\displaystyle H}{|}}{\underset{\underset{\displaystyle H}{|}}{C}}-\overset{..}{\underset{..}{O}}-H \quad \text{and} \quad H-\overset{\overset{\displaystyle H}{|}}{\underset{\underset{\displaystyle H}{|}}{C}}-\overset{..}{\underset{..}{O}}-\overset{\overset{\displaystyle H}{|}}{\underset{\underset{\displaystyle H}{|}}{C}}-H$$

The two molecules are called isomers of each other, and the phenomenon is called isomerism. Although the molecular formulas of both substances are the same, C_2H_6O, their properties differ markedly because of their different atomic arrangements.

Isomerism is very common, particularly in organic chemistry, and when double bonds are present, isomerism can occur in very small molecules:

The first two isomers result from the fact that there is no rotation around a double bond, although such rotation can occur around single bonds. The third isomeric structure cannot be converted to either of the first two without breaking bonds.

With certain molecules, given a fixed atomic geometry, it is possible to satisfy the octet rule with more than one bonding arrangement. The classic example is benzene, whose molecular formula is C_6H_6:

These two structures are called resonance structures, and molecules such as benzene, which have two or more resonance structures, are said to exhibit resonance. The actual bonding in such molecules is thought to be an average of the bonding present in the resonance structures. The stability of molecules exhibiting resonance is found to be higher than that anticipated for any single resonance structure.

Although the conclusions we have drawn regarding molecular geometry and polarity can be obtained from Lewis structures, it is much easier to draw such conclusions from models of molecules and ions. The rules we have cited for octet rule structures transfer readily to models. In many ways the models are easier to construct than are the drawings of Lewis structures on paper. In addition, the models are three-dimensional and hence much more representative of the actual species. Using the models, it is relatively easy to see both geometry and polarity, as well as to deduce Lewis structures. In this experiment you will assemble models for a sizeable number of common chemical species and interpret them in the ways we have discussed.

EXPERIMENTAL PROCEDURE

In this experiment you may work in pairs during the first portion of the laboratory period. The models you will use consist of drilled wooden balls, short sticks, and springs. The balls represent atomic nuclei surrounded by the inner electron shells. The sticks and springs represent electron pairs and fit in the holes in the wooden balls. The model (molecule or ion) consists of wooden balls (atoms) connected by sticks or springs (chemical bonds). Some sticks may be connected to only one atom (nonbonding pairs).

In this experiment we will deal with atoms that obey the octet rule; such atoms have four electron pairs around the central core and will be represented by balls with four tetrahedral holes in which there are four sticks or springs. The only exception will be hydrogen atoms, which share two electrons in covalent compounds, and which will be represented by balls with a single hole in which there is a single stick.

In assembling a molecular model of the kind we are considering, it is possible, indeed desirable, to proceed in a systematic manner. We will illustrate the recommended procedure by developing a model for a molecule with the formula CH_2O.

1. Determine the total number of valence electrons in the species. This is easily done once you realize that the number of valence electrons on an atom is equal to the number of the group to which the atom belongs in the Periodic Table. For CH_2O,

C   Group 4   H   Group 1   O   Group 6

Therefore each carbon atom in a molecule or ion contributes four electrons, each hydrogen atom one electron, and each oxygen atom six electrons. The total number of valence electrons equals the sum of the valence electrons on all of the atoms in the species being studied. For CH_2O this total would be $4 + (2 \times 1) + 6$, or 12 valence electrons. If we are working with an ion, we add one electron for each negative charge or subtract one for each positive charge on the ion.

2. Select wooden balls and sticks to represent the atoms and electron pairs in the molecule. You should use four-holed balls for the carbon atom and the oxygen atom, and one-holed balls to represent the hydrogen atoms. Since there are 12 valence electrons in the molecule and electrons occur in pairs, you will need six sticks to represent the six electron pairs. The sticks will serve both as bonds between atoms and as nonbonding electron pairs.

3. Connect the balls with some of the sticks. (Assemble a skeleton structure for the molecule, joining atoms by single bonds.) In some cases this can only be done in one way. Usually, however, there are various possibilities, some of which are more reasonable than others. In CH_2O the model can be assembled by connecting the two H atom balls to the C atom ball with two of the available sticks, and then using a third stick to connect the C atom and O atom balls.

4. The next step is to use the sticks that are left over in such a way as to fill all the remaining holes in the balls. (Distribute the electron pairs so as to give each atom eight electrons and so satisfy the octet rule.) In the model we have assembled, there is one unfilled hole in the C atom ball, three unfilled holes in the O atom ball, and three available sticks. An obvious way to meet the required condition is to use two sticks to fill two of the holes in the O atom ball, and then use two springs instead of two sticks to connect the C atom and O atom balls. The model as completed is shown in Figure 12.1.

5. Interpret the model in terms of the atoms and bonds represented. The sticks and spatial arrangement of the balls will closely correspond to the electronic and atomic arrangement in the molecule. Given our model, we would describe the CH_2O molecule as being planar with single bonds between carbon and hydrogen atoms and a double bond between the C and O atoms. The H—C—H angle is approximately tetrahedral. There are two nonbonding electron pairs on the O atom. Since all bonds are polar and the molecular symmetry does not cancel the polarity in CH_2O, the molecule is polar. The Lewis structure of the molecule is given below:

$$\overset{\cdot\cdot}{\underset{\cdot\cdot}{O}} = C \diagup^{H} \diagdown_{H}$$

(The compound having molecules with the formula CH_2O is well-known and is called formaldehyde. The bonding and structure in CH_2O are as given by the model.)

6. Investigate the possibility of the existence of isomers or resonance structures. It turns out that in the case of CH_2O one can easily construct an isomeric form which obeys the octet rule, in which the central atom is oxygen rather than carbon. It is found that this isomeric form of CH_2O

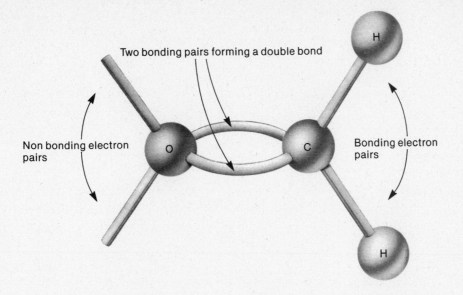

Figure 12.1

does not exist in nature. As a general rule carbon atoms almost always form a total of four bonds; put another way, nonbonding electron pairs on carbon atoms are very rare. Another useful rule of a similar nature is that if a species contains several atoms of one kind and one of another, the atoms of the same kind will assume equivalent positions in the species. In SO_4^{2-}, for example, the four O atoms are all equivalent, and are bonded to the S atom and not to each other.

Resonance structures are reasonably common. For resonance to occur, however, the atomic arrangement must remain fixed for two or more possible electronic structures. For CH_2O there are no resonance structures.

A. Using the procedure we have outlined, construct and report on models of the molecules and ions listed here and/or other species assigned by your instructor. Draw the complete Lewis structure for each molecule, showing nonbonding as well as bonding electrons.

CH_4	H_3O^+	N_2	C_2H_2
CH_2Cl_2	HF	P_4	SO_2
CH_4O	NH_3	C_2H_4	SO_4^{2-}
H_2O	H_2O_2	$C_2H_2Br_2$	CO_2

B. Assuming that stability requires that each atom obey the octet rule, predict the stability of the following species:

$$PCl_3 \qquad CH_3 \qquad OH \qquad CO$$

C. When you have completed parts A and B, see your laboratory instructor, who will check your results and assign you a set of unknown species. Working now by yourself, assemble models for each species as in the previous section, and report on the geometry and bonding in each of the unknown species on the basis of the model you construct. Also consider and report on the polarity and the likelihood of existence of isomers for each species.

EXP. 12

REPORT: Geometrical Structures of Molecules Using Molecular Models

A.

Species	Lewis structure	Geometry	Polarity	Isomers or Resonance
CH_4				
CH_2Cl_2				
CH_4O				
H_2O				
H_3O^+				
HF				

Species	Lewis structure	Geometry	Polarity	Isomers or Resonance
NH_3				
H_2O_2				
N_2				
P_4				
C_2H_4				
$C_2H_2Br_2$				

Continued on following page

Continued

Species	Lewis structure	Geometry	Polarity	Isomers or Resonance
C_2H_2				
SO_2				

Species	Lewis structure	Geometry	Polarity	Isomers or Resonance
SO_4^{2-}				
CO_2				

B. Stability predicted for PCl_3 _____ CH_3 _____ OH _____ CO _____

C. Unknowns _____ _____

Name _____ Section _____

ADVANCE STUDY ASSIGNMENT: Geometrical Structure of Molecules

You are asked by your instructor to construct a model of the CH_2Cl_2 molecule. Being of a conservative nature, you proceed as directed in the section on Experimental Procedure.

1. First you need to find the number of valence electrons in CH_2Cl_2. The number of valence electrons in an atom of an element is equal to the group number of that element in the Periodic Table.

         C is in Group _____   H is in Group _____   Cl is in Group _____

In CH_2Cl_2 there is a total of _____ valence electrons.

2. The model consists of balls and sticks. What kind of ball should you select for the C atom? _____ the H atoms? _____ the Cl atoms? _____ The electrons in the molecule are paired, and each stick represents an electron pair. How many sticks do you need? _____

3. Assemble a skeleton structure for the molecule, connecting the balls with sticks into one unit. Use the rule that C atoms form four bonds, whereas Cl atoms usually do not. Draw a sketch of the skeleton below:

4. How many sticks did you need to make the skeleton structure? _____ How many sticks are left over? _____ If your model is to obey the octet rule each ball must have four sticks in it (except for hydrogen atom balls, which need only one) (Each atom in an octet rule species is surrounded by 4 pairs of electrons.) How many holes remain to be filled? _____ Fill them with the remaining sticks, which represent nonbonding electron pairs. Draw the complete Lewis structure for CH_2Cl_2 using lines for bonds and pairs of dots for nonbonding electrons.

5. Describe the geometry of the model, which is that of CH_2Cl_2 _____ Is the CH_2Cl_2 molecule polar? _____ Why?

Would you expect CH_2Cl_2 to have any isomeric forms? _____ Explain your reasoning.

13 • Classification of Chemical Substances

Depending on the kind of bonding present in a chemical substance, the substance may be called ionic, molecular, or metallic.

In a solid ionic compound there are ions; the large electrostatic forces between the positively and negatively charged ions are responsible for the bonding which holds these particles together.

In a molecular substance the bonding is caused by the sharing of electrons by atoms. When the stable aggregates resulting from covalent bonding contain relatively small numbers of atoms, they are called molecules. If the aggregates are very large and include essentially all the atoms in a macroscopic particle, the substance is called macromolecular.

Metals are characterized by a kind of bonding in which the electrons are much freer to move than in other kinds of substances. The metallic bond is stable but is probably less localized than other bonds.

The terms ionic, molecular, macromolecular, and metallic are somewhat arbitrary, and some substances have properties that would place them in a borderline category, somewhere intermediate between one group and another. It is useful, however, to consider some of the general characteristics of typical ionic, molecular, and macromolecular, and metallic substances, since many very common substances can be readily assigned to one category or another.

IONIC SUBSTANCES

Ionic substances are all solids at room temperature. They are typically crystalline, but may exist as fine powders as well as clearly defined crystals. While many ionic substances are stable up to their melting points, some decompose on heating. It is very common for an ionic crystal to release loosely bound water of hydration at temperatures below 200°C. Anhydrous (dehydrated) ionic compounds have high melting points, usually above 300°C but below 1000°C. They are not readily volatilized and boil at only very high temperatures (Table 13.1).

When molten, ionic compounds conduct an electric current. In the solid state they do not conduct electricity. The conductivity in the molten liquid is attributed to the freedom of motion of the ions, which arises when the crystal lattice is no longer present.

Ionic substances are frequently but not always appreciably soluble in water. The solutions produced conduct the electric current rather well. The conductivity of a solution of a slightly soluble ionic substance is often several times that of the solvent water. Ionic substances are usually not nearly so soluble in other liquids as they are in water. For a liquid to be a good solvent for ionic compounds it must be highly polar, containing molecules with well-defined positive and negative regions with which the ions can interact.

TABLE 13.1   PHYSICAL PROPERTIES OF SOME REPRESENTATIVE CHEMICAL SUBSTANCES

Substance	M.P., °C	B.P., °C	Solubility Water	Solubility CCl_4	Electrical Conductance	Classification
NaCl	801	1413	Sol	Insol	High in melt and in soln	Ionic
MgO	2800	—	Sl sol	Insol	Low in sat'd soln	Ionic
$CoCl_2$	Sublimes	1049	Sol	Insol	High in soln	Ionic
$CoCl_2 \cdot 6H_2O$	86	Dec	Sol	Insol	High in soln	Ionic hydrate, $-H_2O$ at 110°C
$C_{10}H_8$	70	255	Insol	Sol	Zero in melt	Molecular
C_6H_5COOH	122	249	Sl sol	Sol	Low in sat'd soln	Molecular-ionic
$FeCl_3$	282	315	Sol	Insol	High in soln	Molecular-ionic
SnI_4	144	341	Dec	Sol	~Zero in melt	Molecular
SiO_2	1600	2590	Insol	Insol	Zero in solid	Macromolecular
Fe	1535	3000	Insol	Insol	High in solid	Metallic

Key: Sol = at least 0.1 mole/lit; Sl sol = appreciable solubility but <0.1 mole/lit; Insol = essentially insoluble; Dec = decomposes.

MOLECULAR SUBSTANCES

All gases and essentially all liquids at room temperature are molecular in nature. If the molecular weight of a substance is over about a hundred, it may be a solid at room temperature. The melting points of molecular substances are usually below 300°C; these substances are relatively volatile, but a good many will decompose before they boil. Most molecular substances do not conduct the electric current either when solid or when molten.

Organic compounds, which contain primarily carbon and hydrogen, often in combination with other nonmetals, are essentially molecular in nature. Since there are a great many organic substances, it is true that most substances are molecular. If an organic compound decomposes on heating, the residue is frequently a black carbonaceous material. Reasonably large numbers of inorganic substances are also molecular; those which are solids at room temperature include some of the binary compounds of elements in Groups 4, 5, 6 and 7.

Molecular substances are frequently soluble in at least a few organic solvents, with the solubility being enhanced if the substance and the solvent are similar in molecular structure.

Some molecular compounds are markedly polar, which tends to increase their solubility in water and other polar solvents. Such substances may ionize appreciably in water, or even in the melt, so that they become conductors of electricity. Usually the conductivity is considerably lower than that of an ionic material. Most polar molecular compounds in this category are organic, but a few, including some of the salts of the transition metals, are inorganic.

MACROMOLECULAR SUBSTANCES

Macromolecular substances are all solids at room temperature. They have very high melting points, usually above 1000°C, and low volatility. They are typically very resistant to thermal decomposition. They do not conduct electric current and are often good insulators. They are not soluble in water or any organic solvents. They are frequently chemically inert and may be used as abrasives or refractories.

METALLIC SUBSTANCES

The properties of metals appear to derive mainly from the freedom of movement of their bonding electrons. Metals are good electrical conductors in the solid form, and have characteristic luster and malleability. Most metals are solid at room temperature and have melting points that range from below 0°C to over 2000°C. They are not soluble in water or organic solvents. Some metals are prepared as black powders, which may not appear to be electrical conductors; if such powders are heated, the particles will coalesce to give good electrical conductivity.

EXPERIMENTAL PROCEDURE   WEAR YOUR SAFETY GLASSES WHILE PERFORMING THIS EXPERIMENT

In this experiment you will investigate the properties of several substances with the purpose of determining whether they are ionic, molecular, macromolecular, or metallic. In some cases the classification will be very straightforward. In others you may find that the substance behaves in a way that would not clearly place it in a given category but in some intermediate group.

You may use any tests you wish to on the substances, but use due caution when using materials with which you are not familiar. It is suggested that approximate melting point, solubility in water or organic solvents, electrical conductivity of the solid, liquid, or solution, and tendency to decompose may readily be determined and might aid in classification.

Approximate melting points of substances can be determined rather easily. Substances with low melting points, less than 100°C, for example, will melt readily when warmed gently in a test tube. A test tube heated to about 300°C will impart a yellow-orange color to the Bunsen flame. This color becomes more pronounced between 300° and 550°C, at which temperature the Pyrex tube will begin to soften. When heating samples you should *loosely* stopper the test tube with a cork. *Do not breathe* any vapors that are given off. Look for water condensing on the cooler portions of the tube and for indications that sublimation is occurring. For the highest temperature studies possible in the lab, heat the sample in a nickel crucible with a strong Bunsen flame; a noticeable red color will appear in the crucible at about 600°C. If this temperature cannot be achieved with a Bunsen burner, use a Meker burner or a gas-air torch. *Do not heat samples to 600°C* unless their solubility and conductivity properties have been studied at lower temperatures with indecisive results.

Electrical conductivities of your solutions or melts will be measured for you by your laboratory supervisor, who has a portable test meter for that purpose. Distinguish between completely nonconducting, slightly conducting, and highly conducting liquids.

The substances to be studied in the first part of the experiment are on the laboratory tables along with two organic solvents, one polar and one nonpolar. Carry out enough tests on each substance to establish its classification as best you can. Report your observations on each substance, how you would classify it, and your reason for the classification.

When you have completed your tests, report to your laboratory supervisor, who will check your results and issue you two unknowns for characterization.

EXP. 13

OBSERVATIONS AND CONCLUSIONS: Classification of Chemical Substances

Substance No.	Approximate Melting Point, °C ($<$100, 100–300, 300–600, $>$600)	Solubility			Conductivity		Classification and Reason
		H$_2$O	Nonpolar Organic	Polar Organic	Solid Melt	Solution in H$_2$O	
I							
II							
III							
IV							
V							
VI							
Unknown no. _____							

ADVANCE STUDY ASSIGNMENT: Classification of Substances

1. List the properties of a substance which would definitely establish that the material is molecular.

2. If we classify substances as ionic, molecular, macromolecular, or metallic, in which if any categories are all the members

   (a) soluble in water?

   (b) electrical conductors in the melt?

   (c) insoluble in all common solvents?

   (d) solids at room temperature?

3. A given substance is a white solid at 25°C. It melts at 350°C without decomposing and the melt conducts an electric current. What would be the classification of the substance, based on this information?

4. A white solid melts at 1000°C. The melt does not conduct electricity. Classify the substance as best you can from these properties.

EXPERIMENT

14 • Vapor Pressure and Heat of Vaporization of Liquids

The vapor pressure of a pure liquid is the total pressure at equilibrium in a container in which only the liquid and its vapor are present. In a container in which the liquid and other gas are both present, the vapor pressure of the liquid is equal to the partial pressure of its vapor in the container. In this experiment you will measure the vapor pressure of a liquid by determining the increase that occurs in the pressure in a closed container filled with air when the liquid is injected into it.

The vapor pressure of a liquid rises rapidly as the temperature is increased and reaches one atmosphere at the normal boiling point of the liquid. Thermodynamic arguments show that the vapor pressure of a liquid depends on temperature according to the equation:

$$\log_{10} VP = -\frac{\Delta H_{vap}}{2.3RT} + C \tag{1}$$

where VP is the vapor pressure, ΔH_{vap} is the amount of heat in joules required to vaporize one mole of the liquid against a constant pressure, R is the gas constant, 8.31 J/mole K, and T is the absolute temperature. You will note that this equation is of the form

$$Y = BX + C \tag{2}$$

where $Y = \log_{10} VP$, $X = 1/T$ and $B = -\Delta H_{vap}/2.3R$. Consequently, if we measure the vapor pressure of a liquid at various temperatures and plot $\log_{10} VP$ vs. $1/T$, we should obtain a straight line. From the slope B of this line, we can calculate the heat of vaporization of the liquid, since $\Delta H_{vap} = -2.3RB$.

In the laboratory you will measure the vapor pressure of an unknown liquid at approximately 0°C, 20°C, and 40°C, as well as its boiling point at atmospheric pressure. Given the three vapor pressures, you will be able to calculate the heat of vaporization of the liquid by making a graph of $\log_{10} VP$ vs. $1/T$. The graph will then be used to predict the boiling point of the liquid, and the value obtained will be compared with that you found experimentally.

EXPERIMENTAL PROCEDURE

WEAR YOUR SAFETY GLASSES WHILE PERFORMING THIS EXPERIMENT

From the stockroom obtain a suction flask, a rubber stopper fitted with a small dropper, and a short length of rubber tubing, all in a plastic bucket. Also obtain a sample of an unknown liquid.

1. Assemble the apparatus, using the mercury manometer at your lab bench, as indicated in Figure 14.1. The flask should be dry on the inside. If it is not, rinse it with a few ml of acetone (flammable) and blow compressed air into it for a few moments until it is dry. Put a piece of folded rectangular filter paper into the flask, so that liquid from the dropper will fall on the paper. This will facilitate the diffusion of the vapor. Pour some tap water into a beaker and bring it to about 20°C by adding some cold or warm water. Pour this water into the large beaker so that the level of water reaches as far as possible up the neck of the flask. Wait several minutes to ensure that the flask and air inside it are at the temperature of the water. Then remove the stopper from the flask. Pour a small amount of the unknown liquid into a small beaker, and draw about 2 ml of the liquid up into the dropper. Blot any excess liquid from the end of the dropper with a paper towel. Press

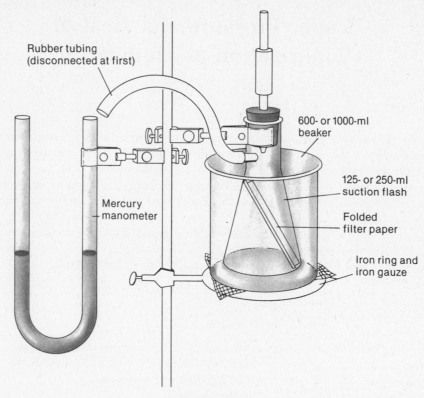

Rubber tubing
(disconnected at first)

600- or 1000-ml
beaker

Mercury
manometer

125- or 250-ml
suction flask

Folded
filter paper

Iron ring and
iron gauze

Figure 14.1

the stopper *firmly* into the flask and connect the hose to the manometer. The mercury levels in the manometer should remain essentially equal.

Immediately squeeze the liquid from the dropper onto the paper, where it will vaporize, diffuse, and exert its vapor pressure. This vapor pressure will be equal to the increase in gas pressure at equilibrium in the container. If you do not observe any appreciable (> 10 mm Hg) pressure increase within a minute or two after injecting the liquid, you probably have a leak in your apparatus and should consult your instructor. When the pressure in the flask becomes steady, in about 10 minutes, read and record the heights of the mercury levels in the manometer to ± 1 mm. An easy way to do this is to measure the height of each level above the lab bench top, using a ruler or meter stick. The *difference in height* is equal to the vapor pressure of the liquid in mm Hg at the temperature of the water bath. Record that temperature to $\pm 0.2°C$.

In this experiment it is essential that (1) the stopper is pressed firmly into the flask, so that the flask plus tubing plus manometer constitute a gas-tight system; (2) no liquid falls into the flask until the manometer is connected; (3) the operations of stoppering the flask, connecting the hose to the manometer, and squeezing the liquid into the flask are conducted with dispatch; (4) you take care when connecting the hose to the manometer, since mercury has a poisonous vapor and is not easily cleaned up when spilled.

Remove the flask from the system. Take out the filter paper. Dry the flask with compressed air, and dry the dropper by squeezing it several times in air.

2.   Starting again at 1, carry out the same experiment at about 0°C. Put a new piece of filter paper into the flask. Use an ice-water bath for cooling the flask. Measure the bath temperature, rather than assuming it to be 0°C.

3.   Repeat the experiment once again, this time holding the water bath at 40°C by judicious heating with a Bunsen burner. Between each run, the flask and dropper must be thoroughly dried, and the precautions noted carefully observed.

4.   In the last part of the experiment you will measure the boiling point of the liquid at the barometric pressure in the laboratory. This is done by pouring the remaining sample of unknown into a large test tube. Determine the boiling point of the liquid using the apparatus shown in

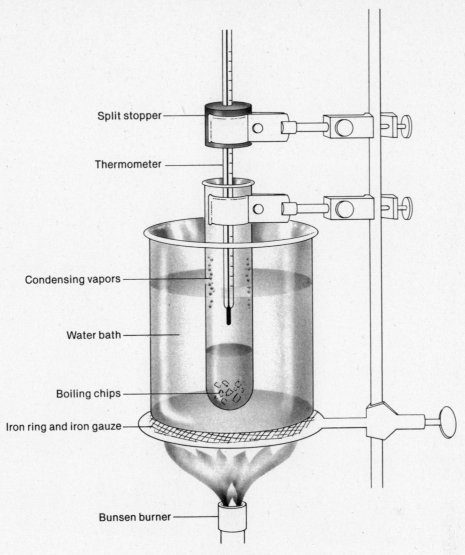

Figure 14.2

Figure 14.2. The thermometer bulb should be just above the liquid surface. Heat the water bath until the liquid in the tube boils gently, with its vapor condensing *at least 5 cm below* the top of the test tube. Boiling chips may help in keeping the liquid boiling smoothly. As the boiling proceeds there will be some condensation on the thermometer and droplets will be falling from the thermometer bulb. After a minute or two the temperature should become reasonably steady at the boiling point of the liquid. Record the temperature under these conditions, along with the barometric pressure in the laboratory. The liquids used may be flammable and toxic, so you should not inhale their vapors unnecessarily. *Do not* heat the water bath so strongly that condensation of the vapors from the liquid occurs only at the top of the test tube.

CAUTION: Mercury vapor is toxic. Use care when working with the mercury manometer. Keep the plastic bucket under the manometer to catch the mercury in case of a spill.

DATA AND CALCULATIONS: Vapor Pressure and Heat of Vaporization
of Liquids

Temperature, t, in °C	Heights of manometer mercury levels in mm		Vapor pressure mm Hg
1. _____	_____	_____	_____
2. _____	_____	_____	_____
3. _____	_____	_____	_____

Boiling point _____ °C

Barometric pressure _____ mm Hg

Using the relation (1) between vapor pressure and temperature, we will calculate the molar heat of vaporization of your liquid and its boiling point from the vapor pressure data obtained. It would be useful to first make the calculations indicated in the following table:

Approximate temperature °C	t, actual temperature °C	T, temperature K	$1/T$	Vapor pressure, VP, in mm Hg	$\log_{10} VP$
0	_____	_____	_____	_____	_____
20	_____	_____	_____	_____	_____
40	_____	_____	_____	_____	_____

On the graph paper provided make a graph of $\log_{10} VP$ vs. $1/T$. Let $\log_{10} VP$ be the ordinate and plot $1/T$ on the abscissa. Since $\log_{10} VP$ is, by (1), a linear function of $1/T$, the line obtained should be nearly straight. Find the slope of the line, $\Delta\log_{10} VP/\Delta(1/T)$.

The slope of the line is equal to $-\Delta H_{\text{vaporization}}/2.3R$. Given that $R = 8.31$ joules/mole K, calculate the molar heat of vaporization, ΔH_{vap}, of your liquid.

$$\text{Slope} = \frac{\Delta\log_{10} VP}{\Delta 1/T} = \underline{\hspace{1.5cm}} = \frac{-\Delta H_{\text{vap}}}{2.3R}, \qquad \Delta H_{\text{vap}} = \underline{\hspace{1.5cm}}\text{joules/mole}$$

From the graph it is also possible to find the temperature at which your liquid will have any given vapor pressure. Recalling that a liquid will boil in an open container at the *temperature* at which its *vapor pressure* is equal to the *atmospheric pressure*, predict the boiling point of the liquid at the barometric pressure in the laboratory.

Continued on following page   **107**

$P_{\text{barometric}}$ _____ mm Hg

$\log_{10} P_{\text{barometric}}$ _____

$1/T$ at this pressure _____ K^{-1} (from graph)

T at this pressure _____ K

t at this pressure _____ °C (boiling point predicted)

Boiling point observed experimentally _____ °C

Unknown no. _____

Does your data appear to follow the equation:

$$\log_{10} VP = \frac{-\Delta H_{\text{vap}}}{2.3RT} + C$$

Give your reasoning.

VAPOR PRESSURE AND HEAT OF
VAPORIZATION OF LIQUIDS
(Data and Calculations)

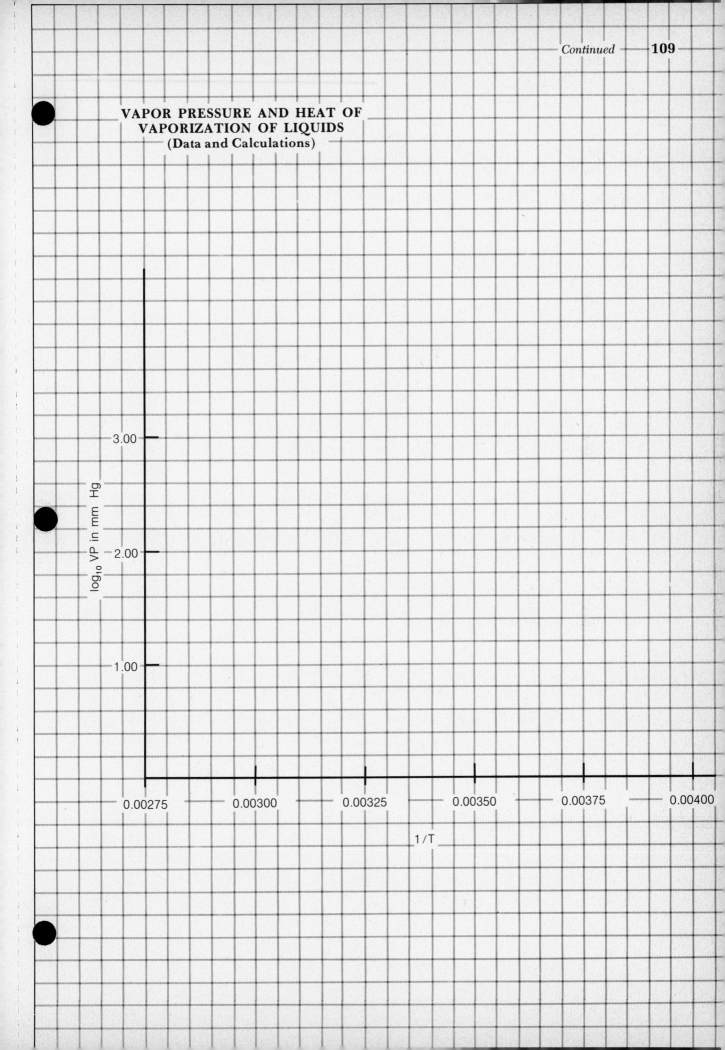

ADVANCE STUDY ASSIGNMENT: Vapor Pressure of Liquids

1. The vapor pressure of benzene was measured by the method of this experiment with the following results (see Fig. 14.1):

$t°C$	Heights of manometer levels in mm Hg Right	Left	Vapor pressure in mm Hg	$\log_{10} VP$	T in K	$1/T$
−2	88	112	_____	_____	_____	_____
19	68	132	_____	_____	_____	_____
35	27	175	_____	_____	_____	_____

  a. Calculate the vapor pressure of benzene at the three temperatures. Find the logarithm to base 10 of each vapor pressure. Enter your results in the table.
  b. Find the absolute temperature T and its reciprocal at each temperature.
  c. Using the graph paper on the next page, plot the three points relating $\log_{10} VP$ to $1/T$. Draw the best straight line you can through these points. (Such a line minimizes the sum of the distances from the points to the line.)
  d. Determine the slope of the straight line. To do this you need to find $\Delta\log_{10} VP/\Delta(1/T)$. Perhaps the easiest way to evaluate these terms is to extend the straight line until it cuts the ordinate (the vertical axis). Take a section of the straight line, say from $1/T$ equals 0.00275 to 0.00375. For that section, $\Delta 1/T$ equals $0.00375 - 0.00275$, or 0.00100. $\Delta\log_{10} VP$ equals the change in $\log_{10} VP$ when you go from the left end of the line to the right ($\Delta\log_{10} VP$ is negative). To obtain the slope, divide $\Delta\log_{10} VP$ by $\Delta(1/T)$.

slope = _____

  e. Noting that $\Delta H_{vap} = -2.3R \times$ slope, calculate the molar heat of vaporization of benzene. (Heats of vaporization of typical liquids are about 30,000 joules/mole.)

$\Delta H_{vap} =$ _____ joules/mole

  f. Benzene will boil when its vapor pressure exceeds the air pressure at its surface. Using the graph from Part c, find $1/T$ when the vapor pressure is 760 mm Hg. Then find T in K and in °C. This temperature is the normal boiling point of benzene.

$1/T =$ _____ ; $T =$ _____ K; $t =$ _____ °C; $t_{obs} = 80°C$

VAPOR PRESSURE OF LIQUIDS
(Advance Study Assignment)

log₁₀ VP in mm Hg axis with values 1.00, 2.00, 3.00

1/T axis with values 0.00275, 0.00300, 0.00325, 0.00350, 0.00375

15 • Molar Mass Determination by Depression of the Freezing Point

In an earlier experiment you observed the change of vapor pressure of a liquid as a function of temperature. If a nonvolatile solid compound (the solute) is dissolved in a liquid, the vapor pressure of the liquid solvent is lowered. This decrease in the vapor pressure of the solvent results in other easily observable physical changes; the boiling point of the solution is higher than that of the pure solvent and the freezing point is lower.

Many years ago chemists observed that at low solute concentrations the changes in the boiling point, the freezing point, and the vapor pressure of a solution are all proportional to the amount of solute that is dissolved in the solvent. These three properties are collectively known as colligative properties of solutions. The colligative properties of a solution depend only on the number of solute particles present in a given amount of solvent and not on the kind of particles dissolved.

When working with boiling point elevations or freezing point depressions of solutions, it is convenient to express the solute concentration in terms of its molality m defined by the relation:

$$\text{molality of } A = m_A = \frac{\text{no. of moles } A \text{ dissolved}}{\text{no. of kg solvent in the solution}}$$

For this unit of concentration, the boiling point elevation, $T_b - T_b^\circ$ or ΔT_b, and the freezing point depression, $T_f^\circ - T_f$ or ΔT_f, in °C at low concentrations are given by the equations:

$$\Delta T_b = k_b m \qquad \Delta T_f = k_f m \qquad (1)$$

where k_b and k_f are characteristic of the solvent used. For water, $k_b = 0.52$ and $k_f = 1.86$. For benzene, $k_b = 2.53$ and $k_f = 5.10$.

One of the main uses of the colligative properties of solutions is in connection with the determination of the molar masses of unknown substances. If we dissolve a known amount of solute in a given amount of solvent and measure ΔT_b or ΔT_f of the solution produced, and if we know the appropriate k for the solvent, we can find the molality and hence the molar mass, MM, of the solute. In the case of the freezing point depression, the relation would be:

$$\Delta T_f = k_f m = k_f \times \frac{\text{no. moles solute}}{\text{no. kg solvent}} = k_f \times \frac{\left(\dfrac{\text{no. g solute}}{\text{MM solute}}\right)}{\text{no. kg solvent}} \qquad (2)$$

In this experiment you will be asked to estimate the molar mass of an unknown solute, using this equation. The solvent used will be paradichlorobenzene, which has a convenient melting point and a relatively large value for k_f, 7.10. The freezing points will be obtained by studying the rate at which liquid paradichlorobenzene and some of its solutions containing the unknown cool in air.

When a pure substance which melts at, say, 60°C is heated to 70°C, where it will be completely liquid, and then allowed to cool in air, the temperature of the sample will vary with time, as in Figure 15.1. Initially the temperature will fall quite rapidly. When the freezing point is reached, solid will begin to form, and the temperature will tend to hold steady until the sample is all solid. The freezing point of the pure liquid is the constant temperature observed while the liquid is freezing to a solid.

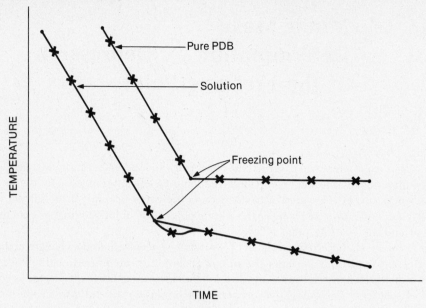

TIME

Figure 15.1

The cooling behavior of a solution is somewhat different from that of a pure liquid, and is also shown in Figure 15.1. The temperature at which the solution begins to freeze is lower than for the pure solvent. In addition, there is a slow gradual fall in temperature as freezing proceeds. The best value for the freezing point of the solution is obtained by drawing two straight lines connecting the points on the temperature-time graph. The first line connects points where the solution is all liquid. The second line connects points where solid and liquid coexist. The point where the two lines intersect is the freezing point of the solution. With both the pure liquid and solutions, at the time when solid first appears, the temperature may fall below the freezing point and then come back up to it as solid forms. The effect is called supercooling, and is shown in Figure 15.1. When drawing the straight line in the solid-liquid region of the graph, ignore points where supercooling was observed. In order to establish the proper straight line in the solid-liquid region it is necessary to record the temperature until the trend with time is smooth and clearly established.

EXPERIMENTAL PROCEDURE

A. Determination of the Freezing Point of Paradichlorobenzene. From the stockroom obtain a stopper fitted with a sensitive thermometer and a glass stirrer, a large test tube, and a sample of solid unknown. **Remember that the thermometer is both fragile and expensive, so handle it with due care.** Weigh the test tube on a top loading or triple beam balance to 0.01 g. Add about 30 g of paradichlorobenzene, PDB, to the test tube and weigh again to the same precision.

Fill your 600 ml beaker almost full of hot water from the faucet. Support the beaker on an iron ring and wire gauze on a ring stand and heat the water with a Bunsen burner. Clamp the test tube to the ring stand and immerse the tube in the water as far as is convenient (see Fig. 15.2). *CAUTION:* The vapor of PDB is toxic, like that of most organic compounds. Avoid breathing the vapor.

When the water gets above the melting point of the PDB, it will begin to melt. After most of the solid appears to have melted, clamp the thermometer-stirrer assembly as in Figure 15.2, adjusting the level of the thermometer so that the bulb is about 1 cm above the bottom of the tube, well down into the melt. Stir to melt the remaining solid PDB. When the PDB is at about 65°C, stop heating. Carefully lower the iron ring and water bath and put the beaker of hot water on the lab bench well away from the test tube. Dry the outside of the test tube with a towel.

Record the temperature of the paradichlorobenzene as it cools in the air. Stir the liquid slowly but continuously to minimize supercooling. Start readings at about 60°C and note the temperature

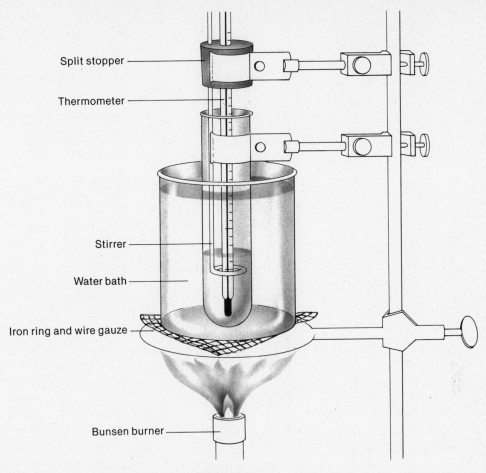

Split stopper

Thermometer

Stirrer

Water bath

Iron ring and wire gauze

Bunsen burner

Figure 15.2

every 30 seconds for 8 minutes or until the liquid has solidified to the point that you are no longer able to stir it. Near the melting point you will begin to observe crystals of PDB in the liquid, and these will increase in amount as cooling proceeds.

B. Determination of the Molar Mass of an Unknown Compound.   Weigh your unknown in its container to 0.01 g. Pour about half of the sample (about 2 g) into your test tube of PDB and reweigh the container. Be careful with both weighings.

Heat the test tube in the water bath until the PDB is again melted and the solid unknown is dissolved. Make sure that all the PDB on the walls is melted; stir well to mix the unknown with the PDB thoroughly. When the melt has reached about 65°C, remove the bath, dry the test tube, and let it cool as before. Start readings at about 58°C and continue to take readings, with stirring, for 8 minutes.

The dependence of temperature on time with the solution will be similar to that observed for pure PDB, except that the first crystals will appear at a lower temperature, and the temperature of the solid-solution system will gradually fall as cooling proceeds. There may be some supercooling, as evidenced by a rise in temperature shortly after the first appearance of PDB crystals.

Add the rest of your unknown (about 2 g) to the PDB solution and again weigh the container. Melt the PDB as before, heating it to about 60°C before removing the water bath. Start temperature-time readings at about 55°C. Repeat the entire procedure described above.

Name _____ Section _____

DATA AND CALCULATIONS: Molar Mass Determination by Freezing Point Depression

Mass of large test tube   _____ g

Mass of test tube plus about 30 g of paradichlorobenzene   _____ g

Mass of container plus unknown   _____ g

Mass of container less Sample I   _____ g

Mass of container less Sample II   _____ g

Time-temperature Readings

Temperature, °C

Time (minutes)	Paradichloro-benzene	Solution I	Solution II
0	_____	_____	_____
½	_____	_____	_____
1	_____	_____	_____
1½	_____	_____	_____
2	_____	_____	_____
2½	_____	_____	_____
3	_____	_____	_____
3½	_____	_____	_____
4	_____	_____	_____
4½	_____	_____	_____
5	_____	_____	_____
5½	_____	_____	_____

Continued on following page

6 _____ _____ _____

$6\frac{1}{2}$ _____ _____ _____

7 _____ _____ _____

$7\frac{1}{2}$ _____ _____ _____

8 _____ _____ _____

Estimation of Freezing Point

On the graph paper provided, plot your temperature vs. time readings for pure paradichloro-benzene and for each run of the two solutions. To avoid overlapping graphs, add 4 minutes to all observed times in making the graph of the cooling curve for pure paradichlorobenzene. Add 2 minutes to all times for Solution I. Use times as observed for Solution II. Connect the points on each cooling curve with two straight lines, ignoring points involving supercooling. The freezing point is at the point where the two lines intersect.

Freezing points (from graphs)

	Pure PDB	Solution I	Solution II
	_____	_____	_____

Calculation of Molar Mass

	Solution I	Solution II
Mass of unknown used (*total* amount in solution)	_____ g	_____ g
Mass of paradichlorobenzene used	_____ g	_____ g
Freezing point of pure paradichlorobenzene	_____ °C	_____ °C
Freezing point of solution	_____ °C	_____ °C
Freezing point depression (total depression)	_____ °C	_____ °C
Total molal concentration of unknown solution (from Eq. 1; $k_f = 7.10$)	_____	_____
Molar mass, MM, of unknown (from Eq. 2)	_____ g	_____ g

Average molar mass _____ g

Unknown no. _____

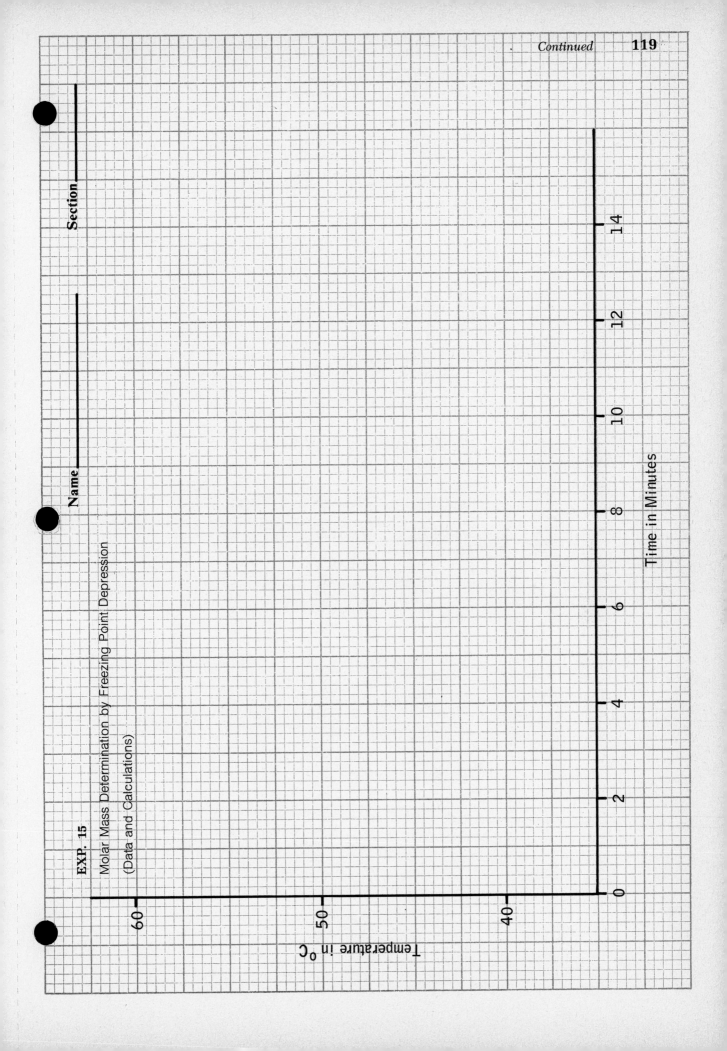

Section

Name

EXP. 15

Molar Mass Determination by Freezing Point Depression

(Data and Calculations)

Temperature in °C

60

50

40

0 2 4 6 8 10 12 14

Time in Minutes

Name _____ Section _____

ADVANCE STUDY ASSIGNMENT: Determination of Molar Mass by
Freezing Point Depression

1. A student determines the freezing point of a solution of 1.92 g of naphthalene in 25.00 g of paradichlorobenzene. He obtains the following temperature-time readings:

Time (min)	0	$\frac{1}{2}$	1	$1\frac{1}{2}$	2	$2\frac{1}{2}$	3	$3\frac{1}{2}$	4
Temperature (°C)	59.7	58.0	56.5	54.8	53.4	52.1	50.9	49.5	48.4

Time (min)	$4\frac{1}{2}$	5	$5\frac{1}{2}$	6	$6\frac{1}{2}$	7	$7\frac{1}{2}$	8
Temperature (°C)	48.6	48.7	48.6	48.5	48.4	48.3	48.2	48.1

a. Plot these data on the graph paper provided. Note that the first several points lie essentially on a straight line. Similarly, the last several points lie on another straight line. Draw in those two lines. The point at which those lines intersect is the freezing point of the solution. (There is a dip in the observed points at about 4 minutes; this is due to supercooling, and is to be ignored when drawing the two lines.)

b. What is the freezing point of the solution?   _____°C

c. What is the freezing point depression, $T_f^\circ - T_f$, or ΔT_f? Take T_f° to   _____°C
be 53.0°C for paradichlorobenzene.

d. What is the molality, m, of the naphthalene? (Use Eq. 1; $k_f = 7.10$)   _____m

e. What is the molar mass of naphthalene? Use Equation 2, as modified below:

$$MM = \frac{\text{no. g solute}}{\text{no. kg solvent}} \times \frac{k_f}{\Delta T_f}$$

_____g

f. The molecular formula of naphthalene is $C_{10}H_8$. Is the
result of Part e consistent with this formula?   _____

Section

Name

EXP. 15

Molar Mass Determination by Freezing Point Depression

(Advance Study Assignment)

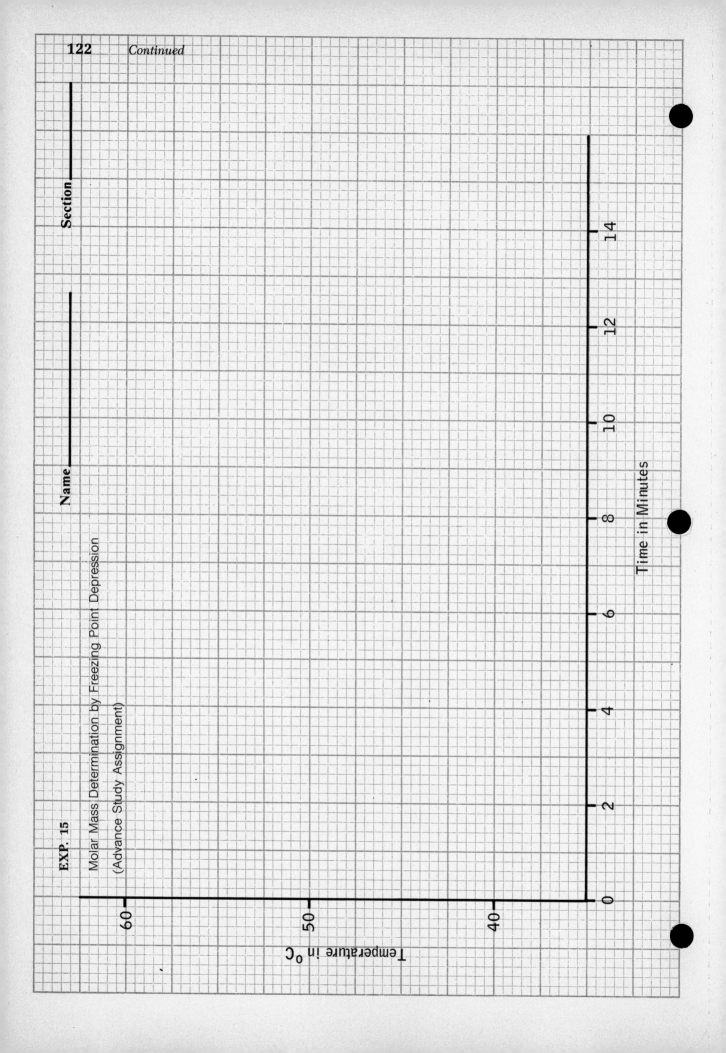

Temperature in °C

60

50

40

0   2   4   6   8   10   12   14

Time in Minutes

EXPERIMENT

16 • Some Nonmetals and their Compounds—Preparation and Properties

Some of the most commonly encountered chemical substances are nonmetallic elements or their simple compounds. O_2 and N_2 in the air, CO_2 produced by combustion of oil, coal, or wood, and H_2O in the rivers, lakes, and air are typical of such substances. Substances containing nonmetallic atoms, whether elementary or compound, are all molecular, reflecting the covalent bonding that holds their atoms together. They are often gases, due to weak intermolecular forces. However, with high molecular masses, hydrogen bonding, or macromolecular structures one finds liquids, like H_2O and Br_2, and solids, like I_2 and graphite.

Several of the common nonmetallic elements and some of their gaseous compounds can be prepared by simple reactions. In this experiment you will prepare some typical examples of such substances and examine a few of their characteristic properties.

EXPERIMENTAL PROCEDURE

WEAR YOUR SAFETY GLASSES WHILE PERFORMING THIS EXPERIMENT

In several of the experiments you will be doing you will prepare gases. In general we will not describe in detail what you should observe, so perform each preparation carefully and report what you actually observe, not what you think we expect you to observe.

There are several tests we will make on the gases you prepare, and the way each of these should be carried out is summarized below.

Test for Odor. To determine the odor of a gas, first pass your hand across the end of the tube, bringing the gas toward your nose. If you don't detect an odor, sniff near the end of the tube, first at some distance and then gradually closer. Don't just put the tube at the end of your nose immediately and take a deep breath. Some of the gases you will make have no odor, and some will have very impressive ones. Some of the gases are very toxic and, even though we will be making only small amounts, caution in testing for odor is important.

Test for Support of Combustion. A few gases will support combustion, but most will not. To make the test, ignite a wood splint with a Bunsen flame, blow out the flame, and put the glowing, but not burning, splint into the gas in the test tube. If the gas supports combustion, the splint will glow more brightly, or may make a small popping noise. If the gas does not support combustion, the splint will go out almost instantly. You can use the same splint for all the support of combustion tests.

Test for Acid-Base Properties. Many gases are acids. This means that if the gas is dissolved in water it will produce some H^+ ions. A few gases are bases; in water solution such gases produce OH^- ions. Other gases do not interact with water, and are neutral. It is easy to establish the acidic or basic nature of a gas by using a chemical indicator. One of the most common acid-base indicators is litmus, which is red in acidic solution and blue in basic solution. To test whether a gas is an acid, moisten a piece of blue litmus paper with water from your wash bottle, and put the paper down into the test tube in which the gas is present. If the gas is an acid, the paper will turn red (blue to red, "as-sed"). Similarly, to test if the gas is a base, moisten a piece of red litmus paper, and hold it down in the tube. A color change to blue will occur if the gas forms a basic solution (red to blue, "alkaloo"). Since you may have used an acid or a base in making the gas, do not touch the walls of the test tube with the paper. The color change will occur fairly quickly and smoothly over the surface of the paper if the gas is acidic or basic. It is not necessary to use a new piece of litmus for each test. Start with a piece of blue and a piece of red litmus. If you need to regenerate the blue paper, hold it over an open bottle of 6 M NH_3, whose vapor is basic. If you need to make red litmus, hold the moist paper over an open bottle of 6 M acetic acid.

A. Preparation and Properties of Nonmetallic Elements: O_2, N_2, Br_2, I_2

1. OXYGEN, O_2. Oxygen can be easily prepared in the laboratory by the decomposition of H_2O_2, hydrogen peroxide, in aqueous solution. Hydrogen peroxide is not very stable and will break down to water and oxygen gas on addition of a suitable catalyst, particularly MnO_2:

$$2 H_2O_2(aq) \xrightarrow{MnO_2} O_2(g) + 2 H_2O \tag{1}$$

Add 1 ml 3% H_2O_2 solution in water to a small test tube. Pick up a very small amount (~0.1 g) of MnO_2 on the tip of your spatula and add it to the liquid in the tube. Hold your finger over the end of the tube to help confine the O_2. Test the evolved gas for odor. Test the gas for any acid-base properties, using moist blue and red litmus paper. Test the gas for support of combustion. Record your observations.

2. NITROGEN, N_2. Sodium sulfamate in water solution will react with nitrite ion to produce nitrogen gas:

$$NO_2^-(aq) + NH_2SO_3^-(aq) \rightarrow N_2(g) + SO_4^{2-}(aq) + H_2O \tag{2}$$

To a small test tube add about 1 ml 1 M KNO_2, potassium nitrite. Add 10–12 drops 0.5 M $NaNH_2SO_3$, sodium sulfamate, and place the test tube in a hot-water bath made from a 250-ml beaker half full of water. Bubbles of nitrogen should form within a few moments. Confine the gas for a few seconds with a stopper. Cautiously test the evolved gas for odor. Carry out the tests for acid-base properties and for support of combustion. If you need to generate more N_2 to complete the tests, add sodium sulfamate solution as necessary, 5–10 drops at a time. Report your observations.

3. IODINE, I_2. The halogen elements are most readily prepared from their sodium or potassium salts. The reaction involves an oxidizing agent, which can remove electrons from the halide ions, freeing the halogen. The reaction which occurs when a solution of potassium iodide, KI, is treated with 6 M HCl and a little MnO_2 is

$$2 I^-(aq) + 4 H^+(aq) + MnO_2(s) \rightarrow I_2(aq) + Mn^{2+}(aq) + 2 H_2O \tag{3}$$

At 25°C I_2 is a solid, with relatively low solubility in water, and appreciable volatility. I_2 can be extracted from aqueous solution into organic solvents, particularly hexane, C_6H_{14} (HEX). The solid, vapor, and solutions in water and HEX all have characteristic colors.

Put 2 drops 1 M KI into a *regular* test tube. Add 6 drops 6M HCl and a tiny amount of manganese dioxide, MnO_2. Swirl the mixture and note any changes which occur. Put the test tube into a hot water bath. After a minute or two a noticeable amount of I_2 vapor should be visible above the liquid. Remove the test tube from the water bath and add 10 ml distilled water. Stopper the tube and shake. Note the color of I_2 in the solution. Sniff the vapor above the solution. Decant the liquid into another regular test tube and add 3 ml hexane, C_6H_{14}. Stopper and shake the tube. Observe the color of the HEX layer and the relative solubility of I_2 in water and HEX. Record your observations.

4. BROMINE, Br_2. Bromine can be made by the same reaction as is used to make iodine, substituting bromide ion for iodide. Bromide ion is less easily oxidized than iodide ion. Bromine at 25°C is a liquid. Br_2 can be extracted from water solution into trichloroethane. Its liquid, vapor, and solutions in water and HEX are colored.

To 2 or 3 drops 1 M NaBr in a regular test tube add 6 drops 6 M HCl and a small amount of

MnO_2, about the size of a small pea. Swirl to mix the reagents, and observe any changes. Heat the tube in the water bath for a minute or two. Try to detect Br_2 vapor above the liquid by observing its color. Add 10 ml water to the tube, stopper, and shake. Note the color of Br_2 in the solution. Sniff the vapor, cautiously. Decant the liquid into a regular test tube. Add 3 ml HEX, stopper and shake. Note the color of Br_2 in HEX. Record your observations.

B. Preparation and Properties of Some Nonmetallic Oxides: CO_2, SO_2, NO, and NO_2

1. **CARBON DIOXIDE, CO_2.**   Carbon dioxide, like several nonmetallic oxides, is easily made by treating an oxyanion with an acid. With carbon dioxide the oxyanion is CO_3^{2-}, carbonate ion, which is present in solutions of carbonate salts, such as Na_2CO_3. The reaction is

$$CO_3^{2-}(aq) + 2\ H^+(aq) \rightarrow H_2CO_3(aq) \rightarrow CO_2(g) + H_2O \tag{4}$$

Carbon dioxide is not very soluble in water, and on acidification, carbonate solutions will tend to effervesce as CO_2 is liberated. Nonmetallic oxides in solution are often acidic but never basic.

To 1 ml 1 M Na_2CO_3 in a small test tube add 6 drops 3 M H_2SO_4. Test the gas for odor, acidic properties, and ability to support combustion.

2. **SULFUR DIOXIDE, SO_2.** Sulfur dioxide is readily prepared by acidification of a solution of sodium sulfite, containing sulfite ion, SO_3^{2-}. As you can see, the reaction which occurs is very similar to that with carbonate ion.

$$SO_3^{2-}(aq) + 2\ H^+(aq) \rightarrow H_2SO_3(aq) \rightarrow SO_2(g) + H_2O \tag{5}$$

Sulfur dioxide is considerably more soluble in water than is carbon dioxide. Some effervescence may be observed on acidification of concentrated sulfite solutions, which increases if the solution is heated in a water bath.

To 1 ml 1 M Na_2SO_3 in a small test tube add 6 drops 3 M H_2SO_4. *Cautiously* test the evolved gas for odor. Test its acidic properties and its ability to support combustion. Put the test tube into the water bath for a few seconds to see if effervescence occurs if the solution is hot.

3. **NITROGEN DIOXIDE, NO_2, AND NITRIC OXIDE, NO.**   If a solution containing nitrite ion, NO_2^-, is treated with acid, two oxides are produced, NO_2 and NO. In solution these gases are combined in the form of N_2O_3, which is colored. When these gases come out of solution, the mixture contains NO and NO_2; the latter is colored. NO is colorless and reacts readily with oxygen in the air to form NO_2. The preparation reaction is

$$2\ NO_2^-(aq) + 2\ H^+(aq) \rightarrow N_2O_3(aq) + H_2O \rightarrow NO(g) + NO_2(g) + H_2O \tag{6}$$

To 1 ml of 1 M KNO_2 in a small test tube, add 6 drops 3 M H_2SO_4. Swirl the mixture for a few seconds and note the color of the solution. Warm the tube in the water bath for a few seconds to increase the rate of gas evolution. Note the color of the gas that is given off. Cautiously test the odor of the gas. Test its acidic properties and its ability to support combustion. Record your observations.

C. Preparation and Properties of Some Nonmetallic Hydrides: NH_3, H_2S

1. **AMMONIA, NH_3.** A 6 M solution of NH_3 in water is a common laboratory reagent. You may have been using it in this experiment to make your litmus paper turn blue. Ammonia gas can be made by simply heating 6 M NH_3. It can also be prepared by addition of a strongly basic solution to a solution of an ammonium salt, such as NH_4Cl. The latter solution contains NH_4^+ ion. On treatment with OH^- ion, as in a solution of NaOH, the following reaction occurs:

$$NH_4^+(aq) + OH^-(aq) \rightarrow NH_3(aq) + H_2O \rightarrow NH_3(g) + H_2O \tag{7}$$

The odor of NH_3 is characteristic. NH_3 is very soluble in water, so effervescence is not observed, even on heating concentrated solutions.

Add 1 ml 1 M NH_4Cl to a small test tube. Add 1 ml 6 M NaOH. Swirl the mixture and test the odor of the evolved gas. Put the test tube into the water bath for a few moments to increase the amount of NH_3 in the gas phase. Test the gas with moistened blue and red litmus paper. Test the gas for support of combustion.

2.   HYDROGEN SULFIDE, H_2S.   Hydrogen sulfide can be made by treating some solid sulfides, particularly FeS, with an acid such as HCl or H_2SO_4. With FeS the reaction is

$$FeS(s) + 2 H^+(aq) \rightarrow H_2S(g) + Fe^{2+}(aq) \tag{8}$$

This reaction was used for many years to make H_2S in the laboratory in courses in qualitative analysis. In this experiment we will employ the method currently used in such courses for H_2S generation. This involves the decomposition of thioacetamide, CH_3CSNH_2, which occurs in solution on treatment with acid and heat. The reaction is

$$CH_3CSNH_2(aq) + 2 H_2O \rightarrow H_2S(g) + CH_3COO^-(aq) + NH_4^+(aq) \tag{9}$$

The odor of H_2S is notorious. H_2S is moderately soluble in water and, since it is produced reasonably slowly in Reaction 9, there will be little if any effervescence.

To 1 ml of 1 M thioacetamide in a small test tube add 6 drops 3 M H_2SO_4. Put the test tube in a boiling water bath for about 1 minute. There may be some cloudiness due to formation of free sulfur. Carefully smell the gas in the tube; H_2S is toxic. Test the gas for acidic and basic properties and for the ability to support combustion.

D.   (Optional).   In this experiment you prepared nine different species containing nonmetallic elements. In each case the source of the species was in solution. In this part of the experiment we will give you an unknown solution which can be used to make one of the nine species. The unknown is one of the solutions you used in this experiment. Identify by suitable tests the species which can be made from your unknown and the substance that is present in your unknown solution.

CAUTION: The following gases prepared in this experiment are toxic: Br_2, SO_2, NO_2, NH_3, and H_2S. Do not inhale these gases unnecessarily when testing their odors. One small sniff will be sufficient and will not be harmful. It is good for you to know these odors, in case you encounter them in the future.

DATA AND OBSERVATIONS: Nonmetals and Their Compounds

A.

Element Prepared	Degree of effervescence	Odor	Acid-base tests	Support of combustion test
1. O_2	_____	_____	_____	_____
2. N_2	_____	_____	_____	_____

	Color		
	Vapor	In H_2O	In TCE
3. I_2	_____	_____	_____
4. Br_2	_____	_____	_____

B.

Oxide Prepared	Degree of effervescence	Odor	Acid test	Color	Support of combustion test
1. CO_2	_____	_____	_____	_____	_____
2. SO_2	_____	_____	_____	_____	_____
3. $NO_2 + NO$	_____	_____	_____	soln _____	_____
				gas _____	

C.

Hydride prepared	Degree of effervescence	Odor	Acid-base tests	Support of combustion test
1. NH_3	_____	_____	_____	_____
2. H_2S	_____	_____	_____	_____

D.   (Optional) Properties of unknown:

Nonmetal species which can be made from unknown _____

Identity of unknown solution _____

Unknown no. _____

Name _____ **Section** _____

ADVANCE STUDY ASSIGNMENT: **Nonmetals and Their Compounds**

   In this experiment nine species are prepared and studied. For each of the species, list the reagents that are used in its preparation and the reaction that occurs.

Reagents used	Reaction

1. O_2

2. N_2

3. I_2

4. Br_2

5. CO_2

6. SO_2

7. $NO_2 + NO$

8. NH_3

9. H_2S

EXPERIMENT

17 • The Solvay Process

In 1979 sodium carbonate, Na_2CO_3, ranked eleventh among industrial chemicals in amount produced in the United States. Production that year was 7.5×10^6 metric tons, about 25 kg for every inhabitant. Na_2CO_3 is used to make glass, soap, paper, and other chemicals. It is one of the so-called heavy chemicals, perhaps because it is made in such large amounts.

Two hundred years ago, before chemistry was a science, sodium carbonate, then called soda ash since it was obtained from wood ashes, was in short supply, and the French government offered a prize for a synthetic method for its manufacture. A scientist named Leblanc developed such a process in 1791, using common salt, limestone, coal, and sulfuric acid. In the process salt was heated with sulfuric acid and converted to salt cake, Na_2SO_4. The salt cake was then heated to about 900°C with a mixture of coal and limestone, producing a mixture of Na_2CO_3, CaS, and various impurities called black ash. The sodium carbonate could be leached from the solid with water and recrystallized. It was a rather messy process, but at least it worked.

The Leblanc process was used to make Na_2CO_3 during much of the nineteenth century. Following its development in 1865, the Solvay process gradually replaced that of Leblanc, and became the source of most of the Na_2CO_3 produced during this century. In 1938 some large deposits of a mineral called trona, $Na_2CO_3 \cdot NaHCO_3 \cdot 2\,H_2O$ were discovered near Green River, Wyoming, and over the past 20 years these deposits have been extensively mined. They now furnish well over half the sodium carbonate needs of this country.

In this experiment we will carry out some of the reactions used in the Solvay process for making Na_2CO_3. The reactants are common salt, NaCl, ammonia, NH_3, and carbon dioxide, CO_2. A moderately concentrated NH_3 solution is first saturated with NaCl. Carbon dioxide is then bubbled through the resulting solution. It reacts with NH_3 and H_2O to form HCO_3^- and NH_4^+ ions. Among the possible ionic products, sodium hydrogen carbonate, $NaHCO_3$, has by far the lowest solubility in water and crystallizes out (solubility in moles/liter at 0°C: $NaHCO_3$, 0.82; NH_4Cl, 5.5; NaCl, 6.1; NH_4HCO_3, 1.5). The solubility of $NaHCO_3$ is lower at low temperatures, so the mixture is cooled to about 0°C to obtain the maximum amount of product. The CO_2 for the process is made by heating limestone, $CaCO_3$, with CaO formed as the other product. The NH_3 used is regenerated from the NH_4^+ ions present in the solution, by reaction with CaO. The final products of the first part of the process are $NaHCO_3$ and $CaCl_2$. Sodium carbonate is then made from $NaHCO_3$ by heating the latter to about 300°C, where it decomposes. The key reactions in the overall Solvay process are summarized below:

$$CaCO_3(s) \xrightarrow{\text{Heat}} CaO(s) + CO_2(g) \tag{1}$$

$$2\,NaCl(s) \rightarrow 2\,Na^+(aq) + 2\,Cl^-(aq) \tag{2}$$

$$2\,NH_3(aq) + 2\,CO_2(aq) + 2\,H_2O \xrightarrow{0°C} 2\,HCO_3^-(aq) + 2\,NH_4^+(aq) \tag{3}$$

$$2\,Na^+(aq) + 2\,HCO_3^-(aq) \xrightarrow{0°C} 2\,NaHCO_3(s) \tag{4}$$

$$CaO(s) + 2\,NH_4^+(aq) \xrightarrow{\text{Heat}} 2\,NH_3(g) + H_2O + Ca^{2+}(aq) \tag{5}$$

$$Ca^{2+}(aq) + 2\,Cl^-(aq) \xrightarrow{\text{Evap.}} CaCl_2(s) \tag{6}$$

$$2\,NaHCO_3(s) \xrightarrow{\text{Heat}} Na_2CO_3(s) + H_2O(g) + CO_2(g) \tag{7}$$

Net reaction (sum of 1–7):

$$CaCO_3(s) + 2\,NaCl(s) \rightarrow Na_2CO_3(s) + CaCl_2(s) \tag{8}$$

The overall reaction in the Solvay process involves exchanging the anions attached to Ca^{2+} and Na^+ but, as you can see, the exchange is accomplished by a very indirect means. The reason we need the indirect means is that Reaction 8 is not spontaneous; rather, the reverse reaction is spontaneous. By carrying out the reaction indirectly, we are able to furnish, at the appropriate places, the energy necessary to make Reaction 8 go. If we just added that energy to the reactants in Equation 8 nothing would happen, at least nothing good.

In our experiment, we will obtain the CO_2 we need from Dry Ice, rather than from $CaCO_3$, since this is a lot more convenient. We will carry out Reactions 2, 3, and 4, but will not attempt to recover the NH_3 nor produce any $CaCl_2$. In the last part of the experiment we will convert $NaHCO_3$ to Na_2CO_3 by Reaction 7.

EXPERIMENTAL PROCEDURE   WEAR YOUR SAFETY GLASSES WHILE PERFORMING THIS EXPERIMENT

I. Preparation of $NaHCO_3$. From the stockroom obtain a 250-ml gas collection bottle and stopper, a large, long 25×200-mm test tube with a stopper assembly, and a length of rubber tubing.

Measure out 40 ml of 3 M NH_3 in a graduated cylinder and pour it into a 250-ml Erlenmeyer flask. Weigh out 15 g NaCl on a rough balance to 0.1 g and add it to the NH_3 solution. Stopper tightly, and shake the solution for several minutes to saturate it with NaCl. If all the NaCl dissolves, add a bit more. When the solution appears to be saturated, filter out the excess solid, using a long-stemmed funnel and filter paper, letting the clear liquid drain into the large, long test tube.

Assemble the apparatus shown in Figure 17.1. The long rubber tubing should be connected to the long bent tube on the stopper assembly. The short straight tube is fitted with a small valve, called a Bunsen valve,° which will vent CO_2 if the pressure becomes excessive. Fill the gas collection bottle about half full with crushed solid CO_2 (Dry Ice). Wrap the bottle with a towel or several paper towels for insulation, and put it into a 600-ml beaker. Insert the stopper firmly in the gas collection bottle.

If your apparatus is tight, you will observe bubbles of CO_2 in the NH_3-NaCl solution; they should be produced at a vigorous rate, and as they bubble up through the solution they will react with NH_3 and become somewhat smaller. As the reaction goes on, the concentration of HCO_3^- ions will gradually increase. While the reaction is proceeding, complete Part I on the data page, and then begin Part II of the experiment. However, check the stoppers occasionally to make sure they haven't popped out.

Finally, after about 30 minutes of bubbling, you should observe a slight cloudiness in the solution as small crystals of $NaHCO_3$ start to form. Continue the bubbling for a half hour more, during which time an appreciable amount of $NaHCO_3$ should form. During the last 15 minutes cool the large test tube in an ice bath, which will decrease the solubility of the $NaHCO_3$ and hence produce more product.

Disconnect the stopper from the large test tube, and unclamp it carefully. Cool the tube in an ice bath for several minutes. Filter out the $NaHCO_3$, using a Buchner funnel with suction. (If you have not yet used a Buchner funnel, see Fig. 2.2.) When all the liquid has been sucked into the suction flask, turn off the suction. Add 3 cc *ice-cold* distilled water to the crystals, wait 10 seconds, and then turn on the suction. This should remove most of the impurities in the $NaHCO_3$. Repeat the procedure with 3 more cc of *ice cold* water. Turn off the suction, and pour 5 ml 95% ethanol onto the crystals. Again, wait about 10 seconds, and then turn on the suction. Repeat with another 5 ml of ethanol. Then leave the suction on for a minute or two to dry the crystals. Turn off the suction, remove the filter paper from the funnel, and transfer the crystals to a piece of weighed filter paper. Weigh the crystals on the paper to 0.1 g.

Place a little of your product into a small test tube. Add a few drops of 6 M HCl. The effervescence of CO_2 is indicative of the presence of the HCO_3^- anion in the solid and shows that the solid is not just NaCl.

° A Bunsen valve consists of a short length of rubber tubing containing a short lengthwise slit. The valve lets gas out, but not in.

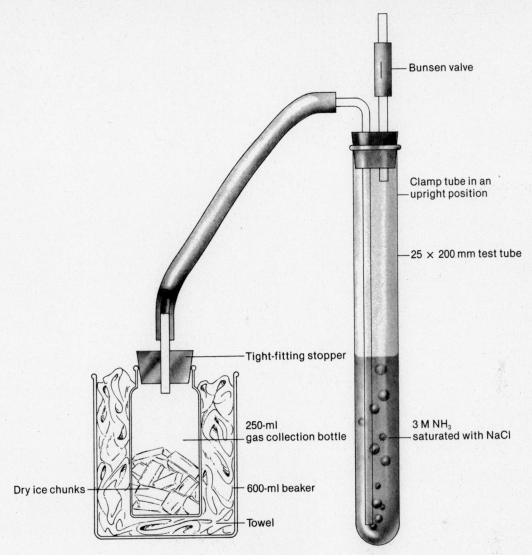

Bunsen valve

Clamp tube in an
upright position

25 × 200 mm test tube

Tight-fitting stopper

250-ml
gas collection bottle

3 M NH₃
saturated with NaCl

Dry ice chunks

600-ml beaker

Towel

Figure 17.1

II. Preparation of Na₂CO₃. Heat a small evaporating dish to redness. Allow it to cool to room temperature, and weigh it on the analytical balance to 0.001 g. Add about one gram of your NaHCO₃ product, which should be dry by now, and weigh again to 0.001 g. Heat the evaporating dish and NaHCO₃ again to redness, holding it at red heat for about 10 minutes. Let the sample cool to room temperature and weigh it.

Rinse out all the borrowed equipment and return it to the stockroom.

DATA AND CALCULATIONS: The Solvay Process

I. Explain why

a. $NaHCO_3$ is the product obtained from the solution made from $NaCl$, NH_3, and CO_2.

b. it takes about a half hour for the first $NaHCO_3$ to appear.

c. one cannot make Na_2CO_3 directly by reaction of $CaCO_3$ and $NaCl$.

II. Mass of filter paper   _____ g

Mass of filter paper + $NaHCO_3$ _____ g

Mass of $NaHCO_3$ formed   _____ g

Mass of evaporating dish   _____ g

Mass of evaporating dish + $NaHCO_3$ _____ g

Mass of evaporating dish + Na_2CO_3 _____ g

Mass of $NaHCO_3$ _____ g

Mass of Na_2CO_3 _____ g

In Reaction 7, $2\,NaHCO_3(s) \xrightarrow{\text{Heat}} Na_2CO_3(s) + H_2O(g) + CO_2(g)$

1. How many grams of Na_2CO_3 would be produced from 1 gram $NaHCO_3$?

_____ g

2. How many grams of Na_2CO_3 were obtained from 1 gram of $NaHCO_3$ in the sample you decomposed?

_____ g

3. How would you explain the difference between the results of 1 and 2?

ADVANCE STUDY ASSIGNMENT: The Solvay Process

1. What is the purpose of the Solvay process?

2. What raw materials are actually used up in the overall process?

3. What are the final products?

4. If 15 g of NaCl were dissolved in 40 ml 3 M NH_3, and if everything went as well as possible in the Solvay process, how much $NaHCO_3$ would crystallize out of the final solution at $0°C$? Assume that NaCl is the limiting reagent, and that the solubility of $NaHCO_3$ is 0.82 moles per liter. Hint: no. moles NaCl = no. moles $NaHCO_3$ produced; find no. moles $NaHCO_3$ remaining in solution and subtract from no. moles produced.

_____ g $NaHCO_3$

5. In Reaction 7, $2\,NaHCO_3(s) \xrightarrow{\text{Heat}} Na_2CO_3(s) + H_2O(g) + CO_2(g)$

    a. How many grams of Na_2CO_3 would be produced from 1.000 gram of $NaHCO_3$?

_____ g Na_2CO_3

    b. How would the mass of Na_2CO_3 per gram $NaHCO_3$ as observed in your experiment vary from the theoretical value if

    (1) the $NaHCO_3$ were wet when it was weighed?

    (2) the $NaHCO_3$ contained NaCl as an impurity?

137

EXPERIMENT

18 • Properties of Systems in Chemical Equilibrium—Le Châtelier's Principle

When working in the laboratory, one often makes observations which at first sight are surprising and hard to explain. One might add a reagent to a solution and obtain a precipitate. Addition of more of that reagent to the precipitate causes it to dissolve. A violet solution turns yellow on addition of a reagent. Subsequent addition of another reagent brings back first a green solution and then the original violet one. Clearly, chemical reactions are occurring, but how and why they behave as they do is not at once obvious.

In this experiment we will examine and attempt to explain several observations of the sort we have mentioned. Central to our explanation will be recognition of the fact that chemical systems tend to exist in a state of equilibrium. If one disturbs the equilibrium in one way or another, the reaction may shift to the left or right, producing the kinds of effects we have mentioned. If one can understand the principles governing the equilibrium system it is often possible to see how one might disturb the system, such as by adding a particular reagent or heat, and so cause it to change in a desirable way.

Before proceeding to specific examples, let us examine the situation in a general way, noting the key principle which allows us to make a system in equilibrium behave as we wish. Consider the reaction

$$A(aq) \rightleftharpoons B(aq) + C(aq) \tag{1}$$

where A, B, and C are molecules or ions in solution. If we have a mixture of these species in equilibrium, it turns out that their concentrations are not completely unrelated. Rather, there is a condition that those concentrations must meet, namely that

$$\frac{[B] \times [C]}{[A]} = K_c \tag{2}$$

where K_c is a constant, called the equilibrium constant for the reaction. For a given reaction at any given temperature K_c has a particular value.

When we say that K_c has a particular value, we mean just that. For example, we might find that, for a given solution in which Reaction 1 can occur, when we substitute the equilibrium values for the molarities of A, B, and C into Equation 2, we get a value of 10 for K_c. Now, suppose we add more of species A to that solution. What will happen? Remember, K_c can't change. If we substitute the new higher molarity of A into Equation 2 we get a value that is smaller than K_c. This means that the system is not in equilibrium, and *must* change in some way to get back to equilibrium. How can it do this? It can do this by shifting to the right, producing more B and C and using up some A. It *must* do this, and *will*, until the molarities of C, B, and A reach values which, on substitution into Equation 2, equal 10. At that point the system is once again in equilibrium. In the new equilibrium state, [B] and [C] are greater than they were initially, and [A] is larger than its initial value but smaller than if there had been no forced shift to the right.

The conclusion you should reach on reading the last paragraph is that *one can always cause a reaction to shift to the right by increasing the concentration of a reactant.* An *increase* in concentration of a *product* will force a *shift* to the *left*. By a similar argument we find that a *decrease* in *reactant* concentration causes a *shift* to the *left*; a *decrease* in *product* concentration produces a *shift* to the *right*. This is all true because K_c does not change (unless you change the temperature). The changes in concentration that one can produce by adding particular reagents may be simply

139

enormous, so the shifts in the equilibrium system may also be enormous. Much of the mystery of chemical behavior disappears once you understand this idea.

Another way one might disturb an equilibrium system is by changing its temperature. When this happens, the value of K_c changes. It turns out that the change in K_c depends upon the enthalpy change, ΔH, for the reaction. If ΔH is positive, greater than zero (endothermic reaction), K_c increases with increasing T. If ΔH is negative (exothermic reaction), K_c decreases with an increase in T. Let us return to our original equilibrium between A, B, and C, where K_c equals 10. Let us assume that ΔH for Reaction 1 is -40 kJ. If we raise the temperature, K_c will go down ($\Delta H < 0$), say to a value of 1. This means that the system will no longer be in equilibrium. Substitution of the initial values of [A], [B], and [C] into Equation 2 produces a value that is too big, 10 instead of 1. How can the system change itself to regain equilibrium? It must of necessity shift to the left, lowering [B] and [C] and raising [A]. This will make the expression in Equation 2 smaller. The shift will continue until the concentrations of A, B, and C, on substitution into Equation 2, give the expression a value of 1.

From the discussion in the previous paragraph, you should be able to conclude that an equilibrium system will shift to the left on being heated if the reaction is exothermic ($\Delta H < 0$, K_c goes down). It will shift to the right if the reaction is endothermic ($\Delta H > 0$, K_c goes up). Again, since we can change temperatures very markedly, we can shift equilibria a long, long way. An endothermic reaction which at $25°C$ has an equilibrium state which consists mainly of reactants might at $1000°C$ exist almost completely as products.

The effects of concentration and temperature on systems in chemical equilibrium are often summarized by Le Châtelier's Principle. The Principle states that:

If you attempt to change a system in chemical equilibrium, it will react in such a way as to counteract the change you attempted.

If you think about the Principle for a while, you will see that it predicts the same kind of behavior as we did by using the properties of K_c. Increasing the concentration of a reactant will, by the Principle, cause a change that decreases that concentration; that change must be a shift to the right. Increasing the temperature of a reaction mixture will cause a change that tends to absorb heat; that change must be a shift in the endothermic direction. The Principle is an interesting one, but does require more careful reasoning in some cases than the more direct approach we employed. For the most part we will find it more useful to base our arguments on the properties of K_c.

In working with aqueous systems, the most important equilibrium is often that which involves the dissociation of water into H^+ and OH^- ions:

$$H_2O \rightleftharpoons H^+(aq) + OH^-(aq) \qquad K_c = [H^+][OH^-] = 1 \times 10^{-14} \qquad (3)$$

In this reaction the concentration of water is very high and is essentially constant at about 55 M; it is incorporated into K_c. The value of K_c is very small, which means that in any water system the product of [H^+] and [OH^-] must be very small. In pure water, [H^+] equals [OH^-] equals 1×10^{-7} M.

Although the product, [H^+] $\times$ [OH^-] is small, that does not mean that both concentrations are necessarily small. If, for example, we dissolve HCl in water, the HCl in the solution will dissociate completely to H^+ and Cl^- ions; in 1 M HCl, [H^+] will become 1 M, and there is nothing that Reaction 3 can do about changing that concentration appreciably. Rather, Reaction 3 must occur in such a direction as to maintain equilibrium. It does this by lowering [OH^-] by reaction to the left; this uses up a little bit of H^+ ion, and drives [OH^-] to the value it must have when [H^+] is 1 M, namely, 1×10^{-14} M. In 1 M HCl, [OH^-] is a factor of ten million *smaller* than it is in water. This makes the properties of 1 M HCl quite different from those of water, particularly where H^+ and OH^- ions are involved.

If we take 1 M HCl and add a solution of NaOH to it, an interesting situation develops. Like HCl, NaOH is completely dissociated in solution, so in 1 M NaOH, [OH^-] is equal to 1 M. If we add 1 M NaOH to 1 M HCl, we will initially raise [OH^-] ions way above 1×10^{-14} M. However, Reaction 3 cannot be in equilibrium when both [H^+] and [OH^-] are high; reaction must occur to reestablish equilibrium. The added OH^- ions react with H^+ ions to form H_2O, decreasing both concentrations until equilibrium is established. If only a small amount of OH^- ion is added, it will essentially all be used up; [H$^+$] will remain high, and [OH$^-$] will still be very small, but somewhat larger than 10^{-14} M. If we add OH$^-$ ion until the amount added equals in moles the amount of H$^+$ originally present, then Reaction 3 will go to the left until [H$^+$] equals [OH$^-$] equals 1×10^{-7} M, and both concentrations will be very small. Further addition of OH$^-$ ion will raise [OH$^-$] to much higher values, easily as high as 1 M. In such a solution, [H$^+$] would be very low, 1×10^{-14} M. So, in aqueous solution, depending on the solutes present, we can have [H$^+$] and [OH$^-$] range from about 1 M to 10^{-14} M, or 14 orders of magnitude. This will have a tremendous effect on any other equilibrium system in which [H$^+$] or [OH$^-$] ions are reactants. Similar situations arise in other equilibrium systems in which the concentration of a reactant or produce can be changed significantly by adding a particular reagent.

In many equilibrium systems, several equilibria are simultaneously present. For example, in aqueous solution, Reaction 3 must always be in equilibrium. There may, in addition, be equilibria between the solutes in the aqueous solution. Some examples are those in Reactions 4, 5, 7, 8, 9, and 10 in the Experimental Procedure section. In some of those reactions, H$^+$ and OH$^-$ ions appear; in others, they do not. In Reaction 4, for example, H$^+$ ion is a product. The molarity of H$^+$ in Reaction 4 is *not* determined by the indicator HMV, since it is only present in a tiny amount. Reaction 4 will have an equilibrium state that is fixed by the state of Reaction 3, which as we have seen depends markedly on the presence of solutes such as HCl or NaOH. Reactions 8 and 9 can similarly be controlled by Reaction 3. Reactions 5 and 7, which do not involve H$^+$ or OH$^-$ ions, are not dependent on Reaction 3 for their equilibrium state. Reaction 10 is sensitive to NH$_3$ concentration and can be driven far to the right by addition of a reagent like 6 M NH$_3$.

EXPERIMENTAL PROCEDURE

WEAR YOUR SAFETY GLASSES WHILE
PERFORMING THIS EXPERIMENT

In this experiment we will work with several equilibrium systems, each of which is similar to the A-B-C system we discussed. We will alter these systems in various ways, forcing shifts to the right and left by changing concentrations and temperature. You will be asked to interpret your observations in terms of the principles we have presented.

A.   Acid-Base Indicators.   There is a large group of chemical substances, called acid-base indicators, which change color in solution when [H$^+$] changes. A typical substance of this sort is called methyl violet, which we will give the formula HMV. In solution HMV dissociates as follows:

$$HMV(aq) \rightleftharpoons H^+(aq) + MV^-(aq) \tag{4}$$
$$\text{yellow} \qquad\qquad\qquad \text{violet}$$

HMV has an intense yellow color, while the anion MV$^-$ is violet. The color of the indicator in solution depends very strongly on [H$^+$].

STEP 1.   Add about 5 ml of distilled water to a regular test tube. Add a few drops of methyl violet indicator. Report the color of the solution on the Data page.

STEP 2.   How could you force the equilibrium system to go to the other form (color)? Select a reagent which should do this and add it to the solution, drop by drop, until the color change is complete. If your reagent works, write its formula on the Data page. If it doesn't, try another until you find one that does. Work with 6 M reagents if they are available.

STEP 3.   Equilibrium systems are reversible. That is, the reaction can be driven to the left and right many times by changing the conditions in the system. How can you force the system in Step 2 to revert to its original color? Select a reagent that should do this and add it drop by drop until the color has become the original one. Again, if your first choice was incorrect, try another reagent. On the Data page write the formula of the reagent that was effective. Answer all the questions for Part A before going on to Part B.

B.   Solubility equilibrium; Finding a value for K_{sp}.   Many ionic substances have limited water solubility. A typical example is $PbCl_2$, which dissolves to some extent in water according to the reaction

$$PbCl_2(s) \rightleftharpoons Pb^{2+}(aq) + 2\ Cl^-(aq) \tag{5}$$

The equilibrium constant for this reaction takes the form

$$K_c = [Pb^{2+}] \times [Cl^-]^2 = K_{sp} \tag{6}$$

The $PbCl_2$ does not enter into the expression because it is a solid, and so has a constant effect on the system, independent of its amount. The equilibrium constant for a solubility equilibrium is called the solubility product, and is given the symbol K_{sp}.

For the equilibrium in Reaction 5 to exist, there *must* be some solid $PbCl_2$ present in the system. If there is no solid, there is no equilibrium; Equation 6 is not obeyed, and $[Pb^{2+}] \times [Cl^-]^2$ must be *less* than the value of K_{sp}. If the solid is present, even in a tiny amount, then the values of $[Pb^{2+}]$ and $[Cl^-]$ are subject to Equation 6.

STEP 1.   Set up a hot water bath, using a 400-ml beaker half full of water. Start heating the water with a burner while proceeding with Step 2.

STEP 2.   To a regular test tube add 5.0 ml 0.3 M $Pb(NO_3)_2$. In this solution $[Pb^{2+}]$ equals 0.3 M. Add 5.0 ml 0.3 M HCl to a 10-ml graduated cylinder. In this solution $[Cl^-]$ equals 0.3 M. Add 1 ml of the HCl solution to the $Pb(NO_3)_2$ solution. Stir, and wait about 15 seconds. What happens? Enter your result.

STEP 3.   To the $Pb(NO_3)_2$ solution add the HCl in 1-ml increments until a noticeable amount of white solid $PbCl_2$ is present after stirring. Record the total volume of HCl added at that point.

STEP 4.   Put the test tube with the precipitate of $PbCl_2$ into the hot water bath. Stir for a few moments. What happens? Enter your observations. Cool the test tube under the cold water tap. What happens?

STEP 5.   Rinse out your graduated cylinder and then add about 5.0 ml of distilled water to the cylinder. Add water in 1-ml increments to the mixture in the test tube, stirring well after each addition. Record the volume of water added when the precipitate just dissolves. Answer the questions and do the calculations in Part B before proceeding.

C.   Complex Ion Equilibria.   Many metallic ions in solution exist not as simple ions but, rather, as complex ions in combination with other ions or molecules, called ligands. For example, the Co^{2+} ion in solution exists as the pink $Co(H_2O)_6^{2+}$ complex ion, and Cu^{2+} as the blue $Cu(H_2O)_4^{2+}$ complex ion. In both of these ions the ligands are H_2O molecules. Complex ions are reasonably stable but may be converted to other complex ions on addition of ligands which form more stable complexes than the original ones. Among the common ligands which may form complex species OH^-, NH_3, and Cl^- are important.

An interesting Co(II) complex is the $CoCl_4^{2-}$ ion, which is blue. This ion is stable in concentrated Cl^- solutions. Depending upon conditions, Co(II) in solution may exist as either $Co(H_2O)_6^{2+}$ or as $CoCl_4^{2-}$. The principles of chemical equilibrium can be used to predict which ion will be present:

$$Co(H_2O)_6^{2+}(aq) + 4\ Cl^-(aq) \rightleftharpoons CoCl_4^{2-}(aq) + 6\ H_2O \tag{7}$$

STEP 1.   Put a few small crystals ($\sim$0.1 g) of $CoCl_2 \cdot 6\ H_2O$ in a regular test tube. Add 2 ml of 12 M HCl (CAUTION) and stir to dissolve the crystals. Record the color of the solution.

STEP 2.   Add 2-ml portions of distilled water, stirring after each dilution, until no further color change occurs. Record the new color.

STEP 3.   Place the test tube into the hot water bath and note any change in color. Cool the tube under the water tap and report your observations.

Complete the questions in Part C before continuing.

D.   Dissolving Insoluble Solids.   We saw in Part B that we can dissolve more $PbCl_2$ by either heating its saturated solution or by simply adding water. In most cases these procedures won't work very well on other solids because they are typically much less soluble than $PbCl_2$.

There are, however, some very powerful methods for dissolving solids that depend for their effectiveness upon the principles of equilibrium. As an example of an insoluble substance, we might consider $Zn(OH)_2$:

$$Zn(OH)_2(s) \rightleftharpoons Zn^{2+}(aq) + 2\,OH^-(aq) \qquad K_{sp} = 5 \times 10^{-17} = [Zn^{2+}] \times [OH^-]^2 \qquad (8)$$

The equilibrium constant for Reaction 8 is very small, which tells us that the reaction does not go very far to the right or, equivalently, that $Zn(OH)_2$ is almost completely insoluble in water. Adding a few drops of a solution containing OH^- ion to one containing Zn^{2+} ion will cause precipitation of $Zn(OH)_2$.

At first sight you might well wonder how one could possibly dissolve, say, 1 mole of $Zn(OH)_2$ in an aqueous solution. If, however, you examine Equation 8, you can see, from the Equation for K_{sp}, that in the saturated solution $[Zn^{2+}] \times [OH^-]^2$ must equal 5×10^{-17}. If, by some means, we can lower that product to a value *below* 5×10^{-17} then $Zn(OH)_2$ will dissolve, until the product becomes equal to K_{sp}, where equilibrium will again exist. To lower the product, we need to lower the concentration of either Zn^{2+} or OH^- very drastically. This turns out to be very easy to do. To lower $[OH^-]$ we can add H^+ ions from an acid. If we do that, we drive Reaction 3 to the left, making $[OH^-]$ very, very small — small enough to dissolve substantial amounts of $Zn(OH)_2$.

To lower $[Zn^{2+}]$ we can take advantage of the fact that zinc(II) forms stable complex ions with both OH^- and NH_3:

$$Zn^{2+}(aq) + 4\,OH^-(aq) \rightleftharpoons Zn(OH)_4^{2-}(aq) \qquad K_1 = 3 \times 10^{15} \qquad (9)$$

$$Zn^{2+}(aq) + 4\,NH_3(aq) \rightleftharpoons Zn(NH_3)_4^{2+}(aq) \qquad K_2 = 1 \times 10^9 \qquad (10)$$

In high concentrations of OH^- ion, Reaction 9 is driven strongly to the right, making $[Zn^{2+}]$ very low. The same thing would happen in solutions containing high concentrations of NH_3. In both media we would therefore expect that $Zn(OH)_2$ might dissolve.

STEP 1.   To each of three small test tubes add about 2 ml 0.1 M $Zn(NO_3)_2$. In this solution $[Zn^{2+}]$ equals 0.1 M. To each test tube add 1 drop 6 M NaOH and stir. Report your observations.

STEP 2.   To the first tube add 6 M HCl drop by drop, with stirring. To the second add 6 M NaOH, again drop by drop. To the third add 6 M NH_3. Note what happens in each case.

STEP 3.   Repeat Steps 1 and 2, this time using a solution of 0.1 M $Mg(NO_3)_2$. Record your observations. Answer the questions in Part D.

DATA AND OBSERVATIONS: Systems in Chemical Equilibrium

A.   1. Color of methyl violet in water _____

2. Reagent causing color change _____ 3. Reagent causing shift back _____
Explain, by considering how changes in $[H^+]$ will cause Reaction 4 to shift to right and left, why the reagents in Steps 2 and 3 caused the solution to change color. Note that Reactions 3 and 4 must both go to equilibrium after a reagent is added.

B.   2. Vol. 0.3 M $Pb(NO_3)_2$ = 5.0 ml   No. moles Pb^{2+} (M × V(lit)) = 1.5×10^{-3} moles

Observations:

3.   Vol. 0.3 M HCl used _____ ml   No. moles Cl^- added _____ moles

4. Observations: in hot water _____   in cold water _____

5. Volume H_2O added to dissolve $PbCl_2$ _____ ml

Total volume of solution _____ ml

    a. Explain why $PbCl_2$ did not precipitate immediately on addition of HCl (What condition must be met by $[Pb^{2+}]$ and $[Cl^-]$ if $PbCl_2$ is to form?)

    b. Explain your observations in Step 4 (In which direction did Reaction 5 shift when heated? What must have happened to the value of K_{sp} in the hot solution? What does this tell you about the sign of ΔH in Reaction 5?)

    c. Explain why the $PbCl_2$ dissolved when water was added in Step 5. (What was the effect of the added water on $[Pb^{2+}]$ and $[Cl^-]$? In what direction would such a change drive Reaction 5?)

Continued on following page   **145**

d. Given the numbers of moles of Pb^{2+} and Cl^- in the final solution in Step 5, and the volume of that solution, calculate $[Pb^{2+}]$ and $[Cl^-]$ in that solution.

$$[Pb^{2+}] \underline{\hspace{3cm}} M; \quad [Cl^-] \underline{\hspace{3cm}} M$$

Noting that the molarities just calculated are essentially those in equilibrium with solid $PbCl_2$, calculate $[Pb^{2+}] \times [Cl^-]^2$. This is equal to K_{sp} for $PbCl_2$.

$$K_{sp} = \underline{\hspace{3cm}}$$

C. 1. Color of $CoCl_2 \cdot 6\,H_2O$ \underline{\hspace{3cm}}

Color in solution in 12 M HCl \underline{\hspace{3cm}}

2. Color in diluted solution \underline{\hspace{3cm}}

3. Color of hot solution \underline{\hspace{3cm}}

Color of cooled solution \underline{\hspace{3cm}}

Formula of Co(II) complex ion present in solution in

a. 12 M HCl \underline{\hspace{3cm}}

b. diluted solution \underline{\hspace{3cm}}

c. hot solution \underline{\hspace{3cm}}

d. cooled solution \underline{\hspace{3cm}}

Explain the color change that occurred when
a. water was added in Step 2. (Consider how a change in $[Cl^-]$ and $[H_2O]$ will shift Reaction 7.)

b. the diluted solution was heated. (How would Reaction 7 shift if K_c went up? How did increasing the temperature affect the value of K_c? What is the sign of ΔH in Reaction 7?)

Continued on following page

D. 1. Observations on addition of a drop 6 M NaOH to $Zn(NO_3)_2$ solution:

2. Effect on solubility of $Zn(OH)_2$:

a. of added HCl soln

b. of added NaOH soln

c. of added NH_3 soln

3. Observations on addition of 1 drop 6 M NaOH to $Mg(NO_3)_2$ solution:

Effect on solubility of $Mg(OH)_2$:

a. of added HCl soln

b. of added NaOH soln

c. of added NH_3 soln

Explain your observations in Steps 1 and 2. (How do increases in $[H^+]$, $[OH^-]$, and $[NH_3]$ affect Reactions 8, 9, and 10, respectively? What is the effect of Reaction 3? How do all these effects influence the solubility of $Zn(OH)_2$?)

In Part 3 you probably found that $Mg(OH)_2$ was similar in some ways to $Zn(OH)_2$ and different in others.

a. How was it similar? Explain that similarity.

b. How was it different? Explain the difference in behavior. (Note that some cations form many complex ions and others do not.)

ADVANCE STUDY ASSIGNMENT: Systems in Chemical Equilibrium

1. Methyl orange, HMO, is a common acid-base indicator. In solution it ionizes according to the equation:

$$HMO(aq) \rightleftarrows H^+(aq) + MO^-(aq)$$
$$\text{red} \qquad\qquad\qquad \text{yellow}$$

If methyl orange is added to distilled water, the solution turns yellow. If a drop or two of 6 M HCl is added to the yellow solution, it turns red. If to that solution one adds a few drops of 6 M NaOH the color reverts to yellow.

    a. Why does adding 6 M HCl to the yellow solution of methyl orange tend to cause the color to change to red? (Note that in solution HCl exists as H^+ and Cl^- ions.)

    b. Why does adding 6 M NaOH to the red solution tend to make it turn back to yellow? (Note that in solution NaOH exists as Na^+ and OH^- ions. How does increasing $[OH^-]$ shift Reaction 3 in the discussion section? How would the resulting change in $[H^+]$ affect the dissociation reaction of HMO?)

2. Magnesium hydroxide is only very slightly soluble in water. The reaction by which it goes into solution is:

$$Mg(OH)_2(s) \rightleftarrows Mg^{2+}(aq) + 2\,OH^-(aq)$$

    a. Formulate the expression for the equilibrium constant, K_{sp}, for the above reaction.

    b. It is possible to dissolve significant amounts of $Mg(OH)_2$ in solutions in which the concentration of either Mg^{2+} or OH^- is very, very small. Explain, using K_{sp}, why this is the case.

    c. Explain why $Mg(OH)_2$ might have very appreciable solubility in 1 M HCl. (Consider the effect of Reaction 3 on the $Mg(OH)_2$ solution reaction.)

EXPERIMENT

19 • Determination of the Equilibrium Constant for a Chemical Reaction

When chemical substances react, the reaction typically does not go to completion. Rather, the system goes to some intermediate state in which both the reactants and products have concentrations which do not change with time. Such a system is said to be in chemical equilibrium. When in equilibrium at a particular temperature, a reaction mixture obeys the Law of Chemical Equilibrium, which imposes a condition on the concentrations of reactants and products. This condition is expressed in the equilibrium constant K_c for the reaction.

In this experiment we will study the equilibrium properties of the reaction between iron(III) ion and thiocyanate ion:

$$Fe^{3+}(aq) + SCN^-(aq) \rightleftharpoons FeSCN^{2+}(aq) \tag{1}$$

When solutions containing Fe^{3+} ion and thiocyanate ion are mixed, Reaction 1 occurs to some extent, forming the $FeSCN^{2+}$ complex ion, which has a deep red color. As a result of the reaction, the equilibrium amounts of Fe^{3+} and SCN^- will be less than they would have been if no reaction occurred; for every mole of $FeSCN^{2+}$ that is formed, one mole of Fe^{3+} and one mole of SCN^- will react. According to the Law of Chemical Equilibrium, the equilibrium constant expression K_c for Reaction 1 is formulated as follows:

$$\frac{[FeSCN^{2+}]}{[Fe^{3+}][SCN^-]} = K_c \tag{2}$$

The value of K_c in Equation 2 is constant at a given temperature. This means that mixtures containing Fe^{3+} and SCN^- will react until Equation 2 is satisfied, so that the same value of the K_c will be obtained no matter what initial amounts of Fe^{3+} and SCN^- were used. Our purpose in this experiment will be to find K_c for this reaction for several mixtures made up in different ways, and to show that K_c indeed has the same value in each of the mixtures. The reaction is a particularly good one to study because K_c is of a convenient magnitude and the color of the $FeSCN^{2+}$ ion makes for an easy analysis of the equilibrium mixture.

The mixtures will be prepared by mixing solutions containing known concentrations of iron(III) nitrate, $Fe(NO_3)_3$, and potassium thiocyanate, $KSCN$. The color of the $FeSCN^{2+}$ ion formed will allow us to determine its equilibrium concentration. Knowing the initial composition of a mixture and the equilibrium concentration of $FeSCN^{2+}$, we can calculate the equilibrium concentrations of the rest of the pertinent species and then determine K_c.

Since the calculations required in this experiment may not be apparent, we will go through a step by step procedure by which they can be made. As a specific example, let us assume that we prepare a mixture by mixing 10.0 ml of 2.00×10^{-3} M $Fe(NO_3)_3$ with 10.0 ml of 2.00×10^{-3} M $KSCN$. As a result of Reaction 1, some red $FeSCN^{2+}$ ion is formed. By the method of analysis described later, its concentration at equilibrium is found to be 1.50×10^{-4} M. Our problem is to find K_c for the reaction from this information. To do this we first need to find the initial number of moles of each reactant in the mixture. Second, we determine the number of moles of product that were formed at equilibrium. Since the product was formed at the expense of reactants, we can calculate the amount of each reactant that was used up. In the third step we find the number of moles of each reactant remaining in the equilibrium mixture. Fourth, we determine the concentration of each reactant. And, finally, in the fifth step, we evaluate K_c for the reaction.

Step 1.   Finding the Initial Number of Moles of Each Reactant.   This requires relating the volumes and concentrations of the reagent solutions that were mixed to the numbers of moles of each reactant species in those solutions. By the definition of the molarity, M_A, of a species A,

$$M_A = \frac{\text{no. moles } A}{\text{no. liters of solution, } V} \quad \text{or no. moles } A = M_A \times V \quad (3)$$

Using Equation 3, it is easy to find the initial number of moles of Fe^{3+} and SCN^-. For each solution the volume used was 10.0 ml, or 0.0100 liter. The molarity of each of the solutions was 2.00×10^{-3} M, so $M_{Fe^{3+}} = 2.00 \times 10^{-3}$ M and $M_{SCN^-} = 2.00 \times 10^{-3}$ M. Therefore, in the reagent solutions, we find that

initial no. moles $Fe^{3+} = M_{Fe^{3+}} \times V = 2.00 \times 10^{-3}$ M $\times$ 0.0100 liter $= 20.0 \times 10^{-6}$ moles

initial no. moles $SCN^- = M_{SCN^-} \times V = 2.00 \times 10^{-3}$ M $\times$ 0.0100 liter $= 20.0 \times 10^{-6}$ moles

Step 2.   Finding the Number of Moles of Product Formed.   Here again we can use Equation 3 to advantage. The concentration of $FeSCN^{2+}$ was found to be 1.50×10^{-4} M at equilibrium. The volume of the mixture at equilibrium is the *sum* of the two volumes that were mixed, and is 20.0 ml, or 0.0200 liter. So,

no. moles $FeSCN^{2+} = M_{FeSCN^{2+}} \times V = 1.50 \times 10^{-4}$ M $\times$ 0.0200 liter $= 3.00 \times 10^{-6}$ moles

The number of moles of Fe^{3+} and SCN^- that were *used up* in producing the $FeSCN^{2+}$ must also both be equal to 3.00×10^{-6} moles since, by Equation 1, it takes *one mole Fe^{3+} and one mole SCN^-* to make each mole of $FeSCN^{2+}$.

Step 3.   Finding the Number of Moles of Each Reactant Present at Equilibrium.   In Step 1 we determined that initially we had 20.0×10^{-6} moles Fe^{3+} and 20.0×10^{-6} moles SCN^- present. In Step 2 we found that in the reaction 3.00×10^{-6} moles Fe^{3+} and 3.00×10^{-6} moles SCN^- were used up. The number of moles present at equilibrium must equal the number we started with minus the number that reacted. Therefore, *at equilibrium,*

no. moles at equilibrium = initial no. moles − no. moles used up

equil. no. moles $Fe^{3+} = 20.0 \times 10^{-6} - 3.00 \times 10^{-6} = 17.0 \times 10^{-6}$ moles   (4)

equil. no. moles $SCN^- = 20.0 \times 10^{-6} - 3.00 \times 10^{-6} = 17.0 \times 10^{-6}$ moles

Step 4.   Find the Concentrations of All Species at Equilibrium.   Experimentally, we obtained the concentration of $FeSCN^{2+}$ directly. $[FeSCN^{2+}] = 1.50 \times 10^{-4}$ M. The concentrations of Fe^{3+} and SCN^- follow from Equation 3. The number of moles of each of these species at equilibrium was obtained in Step 3. The volume of the mixture being studied was 20.0 ml, or 0.0200 liter. So, *at equilibrium,*

$$[Fe^{3+}] = M_{Fe^{3+}} = \frac{\text{no. moles } Fe^{3+}}{\text{volume of solution}} = \frac{17.0 \times 10^{-6} \text{ moles}}{0.0200 \text{ liter}} = 8.5 \times 10^{-4} \text{ M}$$

$$[SCN^-] = M_{SCN^-} = \frac{\text{no. moles } SCN^-}{\text{volume of solution}} = \frac{17.0 \times 10^{-6} \text{ moles}}{0.0200 \text{ liter}} = 8.5 \times 10^{-4} \text{ M}$$

Step 5.   Finding the Value of K_c for the Reaction.   Once the equilibrium concentrations of all the reactants and products are known, one needs merely to substitute into Equation 2 to determine K_c:

$$K_c = \frac{[FeSCN^{2+}]}{[Fe^{3+}][SCN^-]} = \frac{1.50 \times 10^{-4}}{(8.5 \times 10^{-4}) \times (8.5 \times 10^{-4})} = 208$$

In this experiment you will obtain data similar to that shown in this example. The calculations involved in processing that data are completely analogous to those we have made. (Actually, your results will differ from the ones we obtained, since the data in our example were obtained at a different temperature and so relate to a different value of K_c.)

In carrying out this analysis we made the assumption that the reaction which occurred was given by Equation 1. There is no inherent reason why the reaction might not have been

$$Fe^{3+}(aq) + 2\ SCN^-(aq) \rightleftharpoons Fe(SCN)_2^+(aq) \tag{5}$$

If you are interested in matters of this sort, you might ask how we know whether we are actually observing Reaction 1 or Reaction 5. The line of reasoning is the following. If Reaction 1 is occurring, K_c for that reaction as we calculate it should remain constant with different reagent mixtures. If, however, Reaction 5 is the one that is going on, K_c as calculated for that reaction should remain constant. In the optional part of the Calculations section, we will assume that Reaction 5 occurs, and make the analysis of K_c on that basis. The results of the two sets of calculations should make it clear that Reaction 1 is the one that we are studying.

Two analytical methods can be used to determine $[FeSCN^{2+}]$ in the equilibrium mixtures. The more precise method uses a spectrophotometer, which measures the amount of light absorbed by the red complex at 447 nm, the wavelength at which the complex most strongly absorbs. The absorbance, A, of the complex is proportional to its concentration, M, and can be measured directly on the spectrophotometer:

$$A = kM \tag{6}$$

Your instructor will show you how to operate the spectrophotometer, if available to your laboratory, and will provide you with a calibration curve or equation from which you can find $[FeSCN^{2+}]$ once you have determined the absorbance of your solutions.

In the other analytical method a solution of known concentration of $FeSCN^{2+}$ is prepared. The $[FeSCN^{2+}]$ concentrations in the solutions being studied are found by comparing the color intensities of these solutions with that of the known. The method involves matching the color intensity of a given depth of unknown solution with that for an adjusted depth of known solution. The actual procedure and method of calculation are discussed in the Experimental section.

In preparing the mixtures in this experiment we will maintain the concentration of H^+ ion at 0.5 M. The hydrogen ion does not participate directly in the reaction, but its presence is necessary to avoid the formation of brown-colored species such as $FeOH^{2+}$, which would interfere with the analysis of $[FeSCN^{2+}]$.

EXPERIMENTAL PROCEDURE

Label five regular test tubes 1 to 5, with labels or by noting their positions on your test tube rack. Pour about 30 ml 2.00×10^{-3} M $Fe(NO_3)_3$ in 1 M HNO_3 into a dry 100-ml beaker. Pipet 5.0 ml of that solution into each test tube. Then add about 20 ml 2.00×10^{-3} M KSCN to another dry 100-ml beaker. Pipet 1, 2, 3, 4, and 5 ml from the KSCN beaker into each of the corresponding test tubes labeled 1 to 5. Then pipet the proper number of ml of water into each test tube to bring the total volume in each tube to 10.0 ml. The volumes of reagents to be added to each tube are summarized in Table 19.1.

TABLE 19.1

	Test Tube No.				
	1	2	3	4	5
Volume $Fe(NO_3)_3$ solution (ml)	5.0	5.0	5.0	5.0	5.0
Volume KSCN solution (ml)	1.0	2.0	3.0	4.0	5.0
Volume H_2O (ml)	4.0	3.0	2.0	1.0	0.0

Mix each solution thoroughly with a glass stirring rod. Be sure to dry the stirring rod after mixing each solution.

Method I.   Analysis by Spectrophotometric Measurement.   Place a portion of the mixture in tube 1 in a spectrophotometer cell, as demonstrated by your instructor, and measure the absorbance of the solution at 447 nm. Determine the concentration of $FeSCN^{2+}$ from the calibration curve provided for each instrument or from the equation furnished to you. Record the value on the data page. Repeat the measurement using the mixtures in each of the other test tubes.

Method II.   Analysis by Comparison with a Standard.   Prepare a solution of known $[FeSCN^{2+}]$ by pipetting 10.0 ml 0.200 M $Fe(NO_3)_3$ in 1 M HNO_3 into a test tube and adding 2.0 ml 0.002 M KSCN and 8.0 ml water. Mix the solution thoroughly with a stirring rod.

Since in this solution $[Fe^{3+}] \gg [SCN^-]$, Reaction 1 is driven strongly to the right. You can assume without serious error that essentially all the SCN^- added is converted to $FeSCN^{2+}$. Assuming that this is the case, calculate $[FeSCN^{2+}]$ in the standard solution and record the value on the data page.

The $[FeSCN^{2+}]$ in the unknown mixtures in test tubes 1 to 5 can be found by comparing the intensity of the red color in these mixtures with that of the standard solution. This can be done by placing the test tube containing mixture 1 next to a test tube containing the standard. Look down both test tubes toward a well-illuminated piece of white paper on the laboratory bench.

Pour out the standard solution into a dry clean beaker until the color intensity you see down the tube containing the standard matches that which you see when looking down the tube containing the unknown. Use a well-lit piece of white paper as your background. When the colors match, the following relation is valid:

$$[FeSCN^{2+}]_{unknown} \times \text{depth of unknown solution} = [FeSCN^{2+}] \times \text{depth of standard solution} \qquad (7)$$

Measure the depths of the matching solutions with a ruler and record them. Repeat the measurement for Mixtures 2 through 5, recording the depth of each unknown and that of the standard solution which matches it in intensity.

DATA: Determination of the Equilibrium Constant for a Chemical Reaction

Mixture	Volume in ml, 2.00×10^{-3} M $Fe(NO_3)_3$	Volume in ml, 2.00×10^{-3} KSCN	Volume in ml of Water	Method I Absorbance	Method II Depth in mm		$[FeSCN^{2+}]$
					Standard	Unknown	
1	5.0	1.0	4.0	———	———	———	——— $\times 10^{-4}$ M
2	5.0	2.0	3.0	———	———	———	——— $\times 10^{-4}$ M
3	5.0	3.0	2.0	———	———	———	——— $\times 10^{-4}$ M
4	5.0	4.0	1.0	———	———	———	——— $\times 10^{-4}$ M
5	5.0	5.0	0.0	———	———	———	——— $\times 10^{-4}$ M

If Method II was used, $[FeSCN^{2+}]_{standard} =$ ——— $\times 10^{-4}$ M; $[FeSCN^{2+}]$ in Mixtures 1 to 5 found by Equation 7.

CALCULATIONS

A. Calculation of K_c assuming the reaction $Fe^{3+}(aq) + SCN^-(aq) \rightleftharpoons FeSCN^{2+}(aq)$. (1)
This calculation is most easily done by Steps 1 through 5 in the discussion. Results are to be entered in the table on the following page.

Step 1. Find the initial number of moles of Fe^{3+} and SCN^- in the mixtures in test tubes 1 through 5. Use Equation 3 and enter the values in the first two columns in the table.

Step 2. Enter the experimentally determined value of $[FeSCN^{2+}]$ at equilibrium for each of the mixtures in the next to last column in the table. Use Equation 3 to find the number of moles of $FeSCN^{2+}$ in each of the mixtures, and enter the values in the fifth column in the table. Note that this is also the number of moles of Fe^{3+} and SCN^- that were used up in the reaction.

155

Step 3. From the number of moles of Fe^{3+} and SCN^- initially present in each mixture, and the number of moles of Fe^{3+} and SCN^- used up in forming $FeSCN^{2+}$, calculate the number of moles of Fe^{3+} and SCN^- that remain in each mixture at equilibrium. Use Equation 4. Enter the results in Columns 3 and 4 in the table.

Step 4. Use Equation 3 and the results of Step 3 to find the concentrations of all of the species at equilibrium. The volume of the mixture is 10.0 ml, or 0.010 liter in all cases. Enter the values in Columns 6 and 7 in the table.

Step 5. Calculate K_c for the reaction for each of the mixtures by substituting values for the equilibrium concentrations of Fe^{3+}, SCN^-, and $FeSCN^{2+}$ in Equation 2.

Mixture	Initial No. Moles		Equilibrium No. Moles			Equilibrium Concentrations			K_c
	Fe^{3+}	SCN^-	Fe^{3+}	SCN^-	$FeSCN^{2+}$	$[Fe^{3+}]$	$[SCN^-]$	$[FeSCN^{2+}]$	
1	___ $\times 10^{-6}$	___ $\times 10^{-6}$	___ $\times 10^{-6}$	___ $\times 10^{-6}$	___ $\times 10^{-6}$	___ $\times 10^{-4}$	___ $\times 10^{-4}$	___ $\times 10^{-4}$	___
2									
3									
4									
5									

Continued on following page

B. **(Optional)** In calculating K_c in Part A, we assume, correctly, that the formula of the complex ion is $FeSCN^{2+}$. It is by no means obvious that this is the case and one might have assumed, for instance, that $Fe(SCN)_2^+$ was the species formed. The reaction would then be

$$Fe^{3+}(aq) + 2SCN^-(aq) \rightleftharpoons Fe(SCN)_2^+(aq) \qquad (5)$$

If we analyze the equilibrium system we have studied, assuming that Reaction 5 occurs rather than Reaction 1, we would presumably obtain nonconstant values of K_c. Using the same kind of procedure as in Part A, calculate K_c for Mixtures 1, 3, and 5 on the basis that $Fe(SCN)_2^+$ is the formula of the complex ion formed by the reaction between Fe^{3+} and SCN^-. Because of the procedure used for calibrating the system by Method I or Method II, $[Fe(SCN)_2^+]$ will equal *one-half* the $[FeSCN^{2+}]$ obtained for each solution in Part A. Note that *two* moles SCN^- are needed to form *one* mole $Fe(SCN)_2^+$. This changes the expression for K_c. Also, in calculating the equilibrium number of moles SCN^- you will need to subtract $(2 \times$ number of moles $Fe(SCN)_2^+)$ from the initial number of moles SCN^-.

Mixture	Initial No. Moles		Equilibrium No. Moles			Equilibrium Concentrations			K_c
	Fe^{3+}	SCN^-	Fe^{3+}	SCN^-	$Fe(SCN)_2^+$	$[Fe^{3+}]$	$[SCN^-]$	$[Fe(SCN)_2^+]$	
1	___	___	___	___	___	___	___	___	___
3	___	___	___	___	___	___	___	___	___
5	___	___	___	___	___	___	___	___	___

On the basis of the results of Part A, what can you conclude about the validity of the equilibrium concept, as exemplified by Equation 2? What do you conclude about the formula of the iron(III) thiocyanate complex ion?

Name _____ **Section** _____

ADVANCE STUDY ASSIGNMENT: Determination of the Equilibrium Constant for a Reaction

1. A student mixes 5.0 ml 2.00×10^{-3} M $Fe(NO_3)_3$ with 5.0 ml 2.00×10^{-3} M KSCN. She finds that in the equilibrium mixture the concentration of $FeSCN^{2+}$ is 1.4×10^{-4} M. Find K_c for the reaction $Fe^{3+}(aq) + SCN^-(aq) \rightleftharpoons FeSCN^{2+}(aq)$.

Step 1. Find the number of moles Fe^{3+} and SCN^- initially present. (Use Eq. 3).

_____ moles Fe^{3+}; _____ moles SCN^-

Step 2. How many moles $FeSCN^{2+}$ are in the mixture at equilibrium? What is the volume of the equilibrium mixture? (Use Eq. 3.)

_____ ml; _____ moles $FeSCN^{2+}$

How many moles of Fe^{3+} and SCN^- are used up in making the $FeSCN^{2+}$?

_____ moles Fe^{3+}; _____ moles SCN^-

Step 3. How many moles of Fe^{3+} and SCN^- remain in the solution at equilibrium? (Use Eq. 4 and the results of Steps 1 and 2.)

_____ moles Fe^{3+}; _____ moles SCN^-

Step 4. What are the concentrations of Fe^{3+}, SCN^-, and $FeSCN^{2+}$ at equilibrium? What is the volume of the equilibrium mixture? (Use Eq. 3 and the results of Step 3.)

$[Fe^{3+}] =$ _____ M; $[SCN^-] =$ _____ M; $[FeSCN^{2+}] =$ _____ M

_____ ml

Step 5. What is the value of K_c for the reaction? (Use Eq. 2 and the results of Step 4.)

$K_c =$ _____

2. **(Optional)** Assume that the reaction studied in Problem 1 is $Fe^{3+}(aq) + 2\ SCN^-(aq) \rightleftharpoons Fe(SCN)_2^+(aq)$. Find K_c for this reaction, given the data in Problem 1, except that the equilibrium concentration of $Fe(SCN)_2^+$ is equal to 0.7×10^{-4} M.

   a. Formulate the expression for K_c for the alternate reaction just cited.

Continued on following page   **159**

b. Find K_c as you did in Problem 1; take due account of the fact that two moles SCN^- are used up per mole $Fe(SCN)_2^+$ formed.

Step 1.   Results are as in Problem 1.

Step 2.   How many moles of $Fe(SCN)_2^+$ are in the mixture at equilibrium? (Use Eq. 3.)

_____ moles $Fe(SCN)_2^+$

How many moles of Fe^{3+} and SCN^- are used up in making the $Fe(SCN)_2^+$?

_____ moles Fe^{3+}; _____ moles SCN^-

Step 3.   How many moles of Fe^{3+} and SCN^- remain in solution at equilibrium? Use the results of Steps 1 and 2, noting that no. moles SCN^- at equilibrium = original no. moles SCN^- − ($2 \times$ no. moles $Fe(SCN)_2^+$).

_____ moles Fe^{3+}; _____ moles SCN^-

Step 4.   What are the concentrations of Fe^{3+}, SCN^-, and $Fe(SCN)_2^+$ at equilibrium? (Use Eq. 3 and the results of Step 3.)

$[Fe^{3+}]$ = _____ M; $[SCN^-]$ = _____ M; $[Fe(SCN)_2^+]$ = _____ M

Step 5.   Calculate K_c on the basis that the alternate reaction occurs. (Use the answer to Part a.)

K_c = _____

EXPERIMENT

20 • Rates of Chemical Reactions, I. The Iodination of Acetone

The rate at which a chemical reaction occurs depends on several factors: the nature of the reaction, the concentrations of the reactants, the temperature, and the presence of possible catalysts. All of these factors can markedly influence the observed rate of reaction.

Some reactions at a given temperature are very slow indeed; the oxidation of gaseous hydrogen or wood at room temperature would not appreciably proceed in a century. Other reactions are essentially instantaneous; the precipitation of silver chloride when solutions containing silver ions and chloride ions are mixed and the formation of water when acidic and basic solutions are mixed are examples of extremely rapid reactions. In this experiment we will study a reaction which, in the vicinity of room temperature, proceeds at a moderate, relatively easily measured rate.

For a given reaction, the rate typically increases with an increase in the concentration of any reactant. The relation between rate and concentration is a remarkably simple one in many cases, and for the reaction

$$aA + bB \rightarrow cC$$

the rate can usually be expressed by the equation

$$\text{rate} = k(A)^m (B)^n \tag{1}$$

where m and n are generally, but not always, integers, 0, 1, 2, or possibly 3; (A) and (B) are the concentrations of A and B (ordinarily in moles per liter); and k is a constant, called the *rate constant* of the reaction, which makes the relation quantitatively correct. The numbers m and n are called the *orders of the reaction* with respect to A and B. If m is 1 the reaction is said to be *first order* with respect to the reactant A. If n is 2 the reaction is *second order* with respect to reactant B. The overall order is the sum of m and n. In this example the reaction would be *third order* overall.

The rate of a reaction is also significantly dependent on the temperature at which the reaction occurs. An increase in temperature increases the rate, an often cited rule being that a $10°C$ rise in temperature will double the rate. This rule is only approximately correct; nevertheless, it is clear that a rise of temperature of say $100°C$ could change the rate of a reaction appreciably.

As with the concentration, there is a quantitative relation between reaction rate and temperature, but here the relation is somewhat more complicated. This relation is based on the idea that in order to react, the reactant species must have a certain minimum amount of energy present at the time the reactants collide in the reaction step; this amount of energy, which is typically furnished by the kinetic energy of motion of the species present, is called the *activation energy* for the reaction. The equation relating the rate constant k to the absolute temperature T and the activation energy E_a is

$$\log_{10} k = \frac{-E_a}{2.30RT} + \text{constant} \tag{2}$$

where R is the gas constant (8.31 joules/mole K for E_a in joules per mole). This equation is identical in form to Equation 1 in Exp. 14. By measuring k at different temperatures we can determine graphically the activation energy for a reaction.

In this experiment we will study the kinetics of the reaction between iodine and acetone:

$$CH_3-\overset{\overset{\displaystyle O}{\|}}{C}-CH_3(aq) + I_2(aq) \rightarrow CH_3-\overset{\overset{\displaystyle O}{\|}}{C}-CH_2I(aq) + H^+(aq) + I^-(aq)$$

The rate of this reaction is found to depend on the concentration of hydrogen ion in the solution as well as presumably on the concentrations of the two reactants. By Equation 1, the rate law for this reaction is

$$\text{rate} = k(\text{acetone})^m\,(I_2)^n\,(H^+)^p \tag{3}$$

where m, n, and p are the orders of the reaction with respect to acetone, iodine, and hydrogen ion respectively, and k is the rate constant for the reaction.

The rate of this reaction can be expressed as the (small) change in the concentration of I_2, $\Delta(I_2)$, which occurs, divided by the time interval Δt required for the change:

$$\text{rate} = \frac{-\Delta(I_2)}{\Delta t} \tag{4}$$

The minus sign is to make the rate positive ($\Delta(I_2)$ is negative). Ordinarily, since rate varies as the concentrations of the reactants according to Equation 3, in a rate study it would be necessary to measure, directly or indirectly, the concentration of each reactant as a function of time; the rate would typically vary markedly with time, decreasing to very low values as the concentration of at least one reactant becomes very low. This makes reaction rate studies relatively difficult to carry out and introduces mathematical complexities that are difficult for beginning students to understand.

The iodination of acetone is a rather atypical reaction, in that it can be very easily investigated experimentally. First of all, iodine has color, so that one can readily follow changes in iodine concentration visually. A second and very important characteristic of this reaction is that it turns out to be zero order in I_2 concentration. This means (see Equation 3) that the rate of the reaction does not depend on (I_2) at all; $(I_2)^0 = 1$, no matter what the value of (I_2) is, as long as it is not itself zero.

Since the rate of the reaction does not depend on (I_2), we can study the rate by simply making I_2 the limiting reagent present in a large excess of acetone and H^+ ion. We then measure the time required for a known initial concentration of I_2 to be completely used up. If both acetone and H^+ are present at much higher concentrations than that of I_2, their concentrations will not change appreciably during the course of the reaction, and the rate will remain, by Equation 3, effectively constant until all the iodine is gone, at which time the reaction will stop. Under such circumstances, if it takes t seconds for the color of a solution having an initial concentration of I_2 equal to $(I_2)_0$ to disappear, the rate of the reaction, by Equation 4, would be

$$\text{rate} = \frac{-\Delta(I_2)}{\Delta t} = \frac{(I_2)_0}{t} \tag{5}$$

Although the rate of the reaction is constant during its course under the conditions we have set up, we can vary it by changing the initial concentrations of acetone and H^+ ion. If, for example, we should *double* the initial concentration of *acetone* over that in Mixture 1, keeping (H^+) and (I_2) at the *same* values they had previously, then the rate of Mixture 2 would, according to Equation 3, be different from that in Mixture 1:

$$\text{rate } 2 = k(2A)^m\,(I_2)^0\,(H^+)^p \tag{6a}$$
$$\text{rate } 1 = k(A)^m\,(I_2)^0\,(H^+)^p \tag{6b}$$

Dividing the first equation by the second, we see that the k's cancel, as do the terms in the iodine and hydrogen ion concentrations, since they have the same values in both reactions, and we obtain simply

$$\frac{\text{rate } 2}{\text{rate } 1} = \frac{(2A)^m}{(A)^m} = \left(\frac{2A}{A}\right)^m = 2^m \tag{6}$$

Having measured both rate 2 and rate 1 by Equation 5, we can find their ratio, which must be equal to 2^m. We can then solve for m either by inspection or using logarithms and so find the *order* of the reaction with respect to acetone.

By a similar procedure we can measure the order of the reaction with respect to H^+ ion concentration and also confirm the fact that the reaction is zero order with respect to I_2. Having found the order with respect to each reactant, we can then evaluate k, the rate constant for the reaction.

The determination of the orders m and p, the confirmation of the fact that n, the order with respect to I_2, equals zero, and the evaluation of the rate constant k for the reaction at room temperature comprise your assignment in this experiment. You will be furnished with standard solutions of acetone, iodine, and hydrogen ion, and with the composition of one solution that will give a reasonable rate. The rest of the planning and the execution of the experiment will be your responsibility.

An optional part of the experiment is to study the rate of this reaction at different temperatures in order to find its activation energy. The general procedure here would be to study the rate of reaction in one of the mixtures, at room temperature and at two other temperatures, one above and one below room temperature. Knowing the rates, and hence the k's, at the three temperatures, you can then find E_a, the energy of activation for the reaction, by plotting log k vs. $1/T$. The slope of the resultant straight line, by Equation 2, must be $-E_a/2.30R$.

EXPERIMENTAL PROCEDURE

Select two regular test tubes; when filled with distilled water, they should appear to have identical color when you view them down the tubes against a white background.

Draw 50 ml of each of the following solutions into clean dry 100 ml beakers, one solution to a beaker: 4 M acetone, 1 M HCl, and 0.005 M I_2. Cover each beaker with a watch glass.

With your graduated cylinder, measure out 10.0 ml of the 4 M acetone solution and pour it into a clean 125-ml Erlenmeyer flask. Then measure out 10.0 ml 1 M HCl and add that to the acetone in the flask. Add 20.0 ml distilled H_2O to the flask. Drain the graduated cylinder, shaking out any excess water, and then use the cylinder to measure out 10.0 ml 0.005 M I_2 solution. Be careful not to spill the iodine solution on your hands or clothes.

Noting the time on your wrist watch or the wall clock to one second, pour the iodine solution into the Erlenmeyer flask and quickly swirl the flask to thoroughly mix the reagents. The reaction mixture will appear yellow because of the presence of the iodine, and the color will fade slowly as the iodine reacts with the acetone. Fill one of the test tubes $\frac{3}{4}$ full with the reaction mixture, and fill the other test tube to the same depth with distilled water. Look down the test tubes toward a well-lit piece of white paper, and note the time the color of the iodine just disappears. Measure the temperature of the mixture in the test tube.

Repeat the experiment, using as a reference the reacted solution instead of distilled water. The amount of time required in the two runs should agree within about 20 seconds.

The rate of the reaction equals the initial concentration of I_2 *in the reaction mixture* divided by the elapsed time. Since the reaction is zero order in I_2, and since both acetone and H^+ ion are present in great excess, the rate is constant throughout the reaction and the concentrations of both acetone and H^+ remain essentially at their initial values in the reaction mixture.

Having found the reaction rate for one composition of the system, it might be well to think for a moment about what changes in composition you might make to decrease the time and hence increase the rate of reaction. In particular, how could you change the composition in such a way as to allow you to determine how the rate depends upon acetone concentration? If it is not clear how to proceed, reread the discussion preceding Equation 6. In your new mixture you should keep the total volume at 50 ml, and be sure that the concentrations of H^+ and I_2 are the *same* as in the first experiment. Carry out the reaction twice with your new mixture; the times should not differ by more than about 15 seconds. The temperature should be kept within about a degree of that in the initial run. Calculate the rate of the reaction. Compare it with that for the first mixture, and then calculate the order of the reaction with respect to acetone, using a relation similar to Equation 6. First, write an equation like (6a) for the second reaction mixture, substituting in the values for the rate as obtained by Equation 5 and the initial concentration of acetone, I_2, and H^+ in the reaction mixture. Then write an equation like (6b) for the first reaction mixture, using the observed rate and the initial concentrations in that mixture. Obtain an equation like (6) by dividing Equation 6a by Equation 6b. Solve Equation 6 for the order m of the reaction with respect to acetone.

Again change the composition of the reaction mixture so that this time a measurement of the reaction will give you information about the order of the reaction with respect to H^+. Repeat the experiment with this mixture to establish the time of reaction to within 15 seconds, again making sure that the temperature is within about a degree of that observed previously. From the rate you determine for this mixture find p, the order of the reaction with respect to H^+.

Finally, change the reaction mixture composition in such a way as to allow you to show that the order of the reaction with respect to I_2 is zero. Measure the rate of the reaction twice, and calculate n, the order with respect to I_2.

Having found the order of the reaction for each species on which the rate depends, evaluate k, the rate constant for the reaction, from the rate and concentration data in each of the mixtures you studied. If the temperatures at which the reactions were run are all equal to within a degree or two, k should be about the same for each mixture.

(**Optional**) As a final reaction, make up a mixture using reactant volumes that you did not use in any previous experiments. Using Equation 3, the values of concentrations in the mixtures, the orders, and the rate constant you calculated from your experimental data, predict how long it will take for the I_2 color to disappear from your mixture. Measure the time for the reaction and compare it with your prediction.

If time permits, select one of the reaction mixtures you have already used which gave a convenient time, and use that mixture to measure the rate of reaction at about 10°C and at about 40°C. From the two rates you find, plus the rate at room temperature, calculate the energy of activation for the reaction, using Equation 2.

DATA AND CALCULATIONS: The Iodination of Acetone

I. Reaction Rate Data

Mixture	Volume in ml 4 M acetone	Volume in ml 1 M HCl	Volume in ml 0.005 M I_2	Volume in ml H_2O	Time for Reaction in sec 1st Run	2nd Run	Temp °C
I	10	10	10	20	_____	_____	_____
II	_____	_____	_____	_____	_____	_____	_____
III	_____	_____	_____	_____	_____	_____	_____
IV	_____	_____	_____	_____	_____	_____	_____

II. Determination of Reaction Orders with Respect to Acetone, H^+ Ion, and I_2

$$\text{rate} = k\,(\text{acetone})^m\,(I_2)^n\,(H^+)^p \tag{3}$$

Calculate the *initial* concentrations of acetone, H^+ ion, and I_2 in each of the mixtures you studied. Use Equation 5 to find the rate of each reaction.

Mixture	(acetone)	(H^+)	$(I_2)_0$	$\text{rate} = \dfrac{(I_2)_0}{\text{ave. time}}$
I	0.8 M	0.2 M	0.001 M	_____
II	_____	_____	_____	_____
III	_____	_____	_____	_____
IV	_____	_____	_____	_____

Substituting the initial concentrations and the rate from the table above, write Equation 3 as it would apply to Reaction Mixture II:

Rate II =

Now write Equation 3 for Reaction Mixture I, substituting concentrations and the calculated rate from the table:

Rate I =

Divide the equation for Mixture II by the equation for Mixture I; the resulting equation should have the ratio of Rate II to Rate I on the left side, and a ratio of acetone concentrations

Continued on following page   **165**

raised to the power m on the right. It should be similar in appearance to Equation 6. Put the resulting equation below:

$$\frac{\text{rate II}}{\text{rate I}} =$$

The only unknown in the equation is m. Solve for m. $m = $ _____

Now write Equation 3 as it would apply to Reaction Mixture III and as it would apply to Reaction Mixture IV:

$$\text{rate III} =$$

$$\text{rate IV} =$$

Using the ratios of the rates of Mixtures III and IV to those of Mixtures II or I, find the orders of the reaction with respect to H^+ ion and I_2:

$$\frac{\text{rate III}}{\text{rate} _____} =$$ $p = $ _____

$$\frac{\text{rate IV}}{\text{rate} _____} =$$ $n = $ _____

III. Determination of the Rate Constant k

Given the values of m, p, and n as determined in Part II, calculate the rate constant k for each mixture by simply substituting those orders, the initial concentrations, and the observed rate from the table into Equation 3.

Mixture	I	II	III	IV	average
k	_____	_____	_____	_____	_____

IV. (Optional) Prediction of Reaction Rate

Reaction mixture

Volume in ml	Volume in ml	Volume in ml	Volume in ml
4 M acetone _____	1 M HCl _____	0.005 M I_2 _____	H_2O _____

Initial concentrations:

(acetone) _____ M (H^+) _____ M $(I_2)_0$ _____ M

Predicted rate _____ (Equation 3)

Predicted time for reaction _____ sec (Equation 5)

Observed time for reaction _____ sec

Continued on following page

V. (Optional) Determination of Energy of Activation

Reaction mixture used _____(same for all temperatures)

Time for reaction at about 10°C _____ sec   temperature _____°C

Time for reaction at about 40°C _____ sec   temperature _____°C

Time for reaction at room temp. _____ sec   temperature _____°C

Calculate the rate constant at each temperature from your data, following the procedure in III.

	rate	k	$\log k$	$\dfrac{1}{T(K)}$
~10°C	_____	_____	_____	_____
~40°C	_____	_____	_____	_____
room temp.	_____	_____	_____	_____

Plot $\log k$ vs. $1/T$. Find the slope of the best straight line through the points.

Slope = _____

By Equation 2:

$$E_a = -2.30 \times 8.31 \times \text{slope}$$

$E_a =$ _____ joules

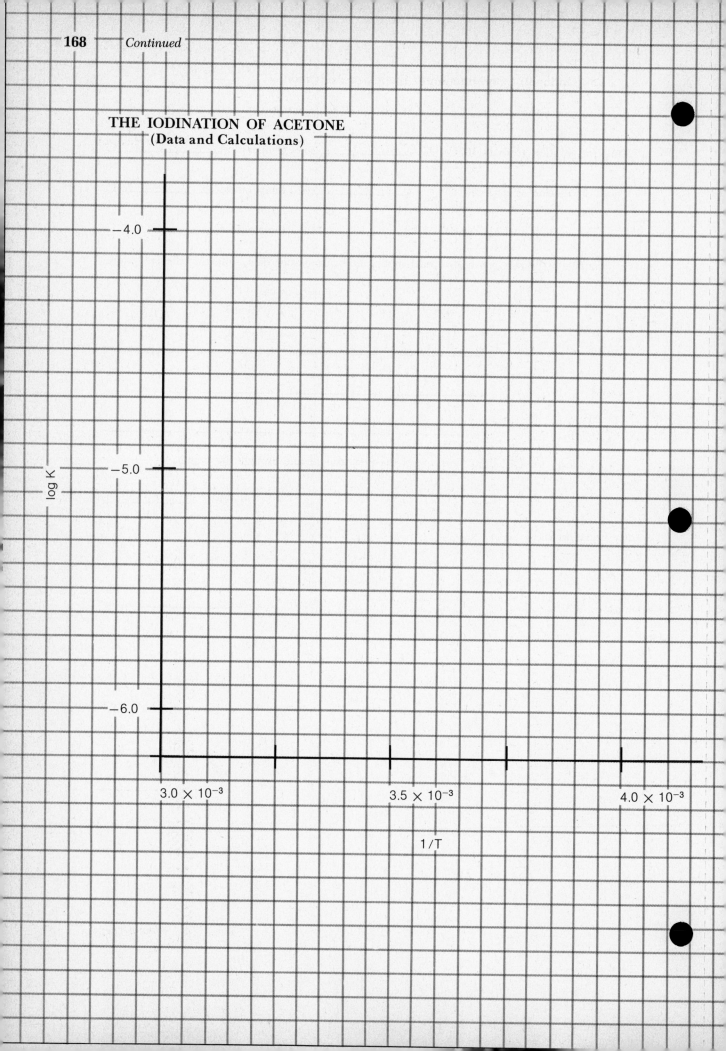

THE IODINATION OF ACETONE
(Data and Calculations)

ADVANCE STUDY ASSIGNMENT: **The Iodination of Acetone**

1. In a reaction involving the iodination of acetone, the following volumes were used to make up the reaction mixture:

$$10 \text{ ml } 4 \text{ M acetone} + 10 \text{ ml } 1 \text{ M HCl} + 10 \text{ ml } 0.005 \text{ M } I_2 + 20 \text{ ml } H_2O$$

   a. How many moles of acetone were in the reaction mixture? Recall that, for a component A, no. moles $A = M_A \times V$, where M_A is the molarity of A and V the volume in liters of the solution of A that was used.

_____ moles acetone

   b. What was the molarity of acetone in the *reaction mixture?* The volume of the *mixture* was 50 ml, 0.050 liter, and the number of moles of acetone was found in Part a. Again, $M_A = \dfrac{\text{no. moles } A}{V \text{ of soln. in liters}}$

_____ M acetone

   c. How could you double the molarity of the acetone in the reaction mixture, keeping the total volume at 50 ml and keeping the same concentrations of H^+ ion and I_2 as in the original mixture?

2. Using the reaction mixture in Problem 1, a student found that it took 250 seconds for the color of the I_2 to disappear.
   a. What was the rate of the reaction? Hint: First find the initial concentration of I_2 in the reaction mixture, $(I_2)_0$. Then use Equation 5.

rate = _____

   b. Given the rate from Part a, and the initial concentrations of acetone, H^+ ion, and I_2 in the reaction mixture, write Equation 3 as it would apply to the mixture.

rate =

Continued on following page   **169**

c. What are the unknowns that remain in the equation in Part b?

—————— —————— —————— ——————

3. A second reaction mixture was made up in the following way:

    20 ml 4 M acetone + 10 ml 1 M HCl + 10 ml 0.005 M I_2 + 10 ml H_2O

   a. What were the initial concentrations of acetone, H^+ ion, and I_2 in the reaction mixture?

       (acetone) ——————— M; (H^+) ——————— M; $(I_2)_0$ ——————— M

   b. It took 120 seconds for the I_2 color to disappear from the reaction mixture when it occurred at the same temperature as the reaction in Problem 2.
      What was the rate of the reaction?   ———————

      Write Equation 3 as it would apply to the second reaction mixture:

      rate =

   c. Divide the equation in Problem 3b by the equation in Problem 2b. The resulting equation should have the ratio of the two rates on the left side and a ratio of acetone concentrations raised to the *m* power on the right. Write the resulting equation and solve for the value of *m*, the order of the reaction with respect to acetone. (Round off the value of *m* to the nearest integer.)

                                              m = ———————

EXPERIMENT

21 • Rates of Chemical Reactions, II. A Clock Reaction

In the previous experiment we discussed the factors that influence the rate of a chemical reaction and presented the terminology used in quantitative relations in studies of the kinetics of chemical reactions. That material is also pertinent to this experiment and should be studied before you proceed further.

This experiment involves the study of the rate properties, or chemical kinetics, of the following reaction between iodide ion and bromate ion under acidic conditions:

$$6\,I^-(aq) + BrO_3^-(aq) + 6\,H^+(aq) \rightarrow 3\,I_2(aq) + Br^-(aq) + 3\,H_2O \tag{1}$$

This reaction proceeds reasonably slowly at room temperature, its rate depending on the concentrations of the I^-, BrO_3^-, and H^+ ions according to the rate law discussed in the previous experiment. For this reaction the rate law takes the form

$$\text{rate} = k(I^-)^m(BrO_3^-)^n(H^+)^p \tag{2}$$

One of the main purposes of the experiment will be to evaluate the rate constant k and the reaction orders m, n, and p for this reaction. We will also investigate the manner in which the reaction rate depends on temperature and will evaluate the activation energy E_a for the reaction.

Our method for measuring the rate of the reaction involves what is frequently called a "clock" reaction. In addition to Reaction 1, whose kinetics we will study, the following reaction will also be made to occur simultaneously in the reaction flask:

$$I_2(aq) + 2\,S_2O_3^{2-}(aq) \rightarrow 2\,I^-(aq) + S_4O_6^{2-}(aq) \tag{3}$$

As compared with (1) this reaction is essentially instantaneous. The I_2 produced in (1) reacts completely with the thiosulfate, $S_2O_3^{2-}$, ion present in the solution, so that until all the thiosulfate ion has reacted, the concentration of I_2 is effectively zero. As soon as the $S_2O_3^{2-}$ is gone from the system, the I_2 produced by (1) remains in the solution and its concentration begins to increase. The presence of I_2 is made strikingly apparent by a starch indicator which is added to the reaction mixture, since I_2 even in small concentrations reacts with starch solution to produce a blue color.

By carrying out Reaction 1 in the presence of $S_2O_3^{2-}$ and a starch indicator, we introduce a "clock" into the system. Our clock tells us when a given amount of BrO_3^- ion has reacted ($\frac{1}{6}$ mole BrO_3^- per mole $S_2O_3^{2-}$), which is just what we need to know, since the rate of reaction can be expressed in terms of the time it takes for a particular amount of BrO_3^- to be used up. In all our reactions, the amount of BrO_3^- that reacts in the time we measure will be constant and small as compared to the amounts of any of the other reactants. This means that the concentrations of all reactants will be essentially constant in Equation 2, and hence so will the rate during each reaction.

In our experiment we will carry out the reaction between BrO_3^-, I^-, and H^+ ions under different concentration conditions. Measured amounts of each of these ions in water solution will be mixed in the presence of a constant small amount of $S_2O_3^{2-}$. The time it takes for each mixture to turn blue will be measured. The time obtained for each reaction will be inversely proportional to its rate. By changing the concentration of one reactant and keeping the other concentrations constant, we can investigate how the rate of the reaction varies with the concentration of a particular reactant. Once we know the order for each reactant we can determine the rate constant for the reaction.

In the last part of the experiment we will investigate how the rate of the reaction depends on temperature. You will recall that in general the rate increases sharply with temperature. By measuring how the rate varies with temperature we can determine the activation energy, E_a, for the reaction by making use of the Arrhenius equation:

$$\log_{10} k = \frac{-E_a}{2.30RT} + \text{constant} \tag{4}$$

In this equation, k is the rate constant at the Kelvin temperature T, E_a is the activation energy, and R is the gas constant. By plotting $\log_{10} k$ against $1/T$ we should obtain, by Equation 4, a straight line whose slope equals $-E_a/2.30R$. From the slope of that line we can easily calculate the activation energy.

EXPERIMENTAL PROCEDURE

A. **Dependence of Reaction Rate on Concentration.**   In Table 21.1 we have summarized the reagent volumes to be used in carrying out the several reactions whose rates we need to know in order to find the general rate law for Reaction 1. First, measure out 100 ml of each of the listed reagents (except H_2O) into clean, labeled flasks or beakers. Use these reagents in your reaction mixtures.

TABLE 21.1   REACTION MIXTURES AT ROOM TEMPERATURE (REAGENT VOLUMES IN ML)

Reaction Mixture	Reaction Flask I (250 ml)			Reaction Flask II (125 ml)	
	0.010 M KI	0.001 M $Na_2S_2O_3$	H_2O	0.040 M $KBrO_3$	0.10 M HCl
1	10	10	10	10	10
2	20	10	0	10	10
3	10	10	0	20	10
4	10	10	0	10	20
5	8	10	12	5	15

The actual procedure for each reaction mixture will be much the same, and we will describe it now for Reaction Mixture 1.

Since there are several reagents to mix, and since we don't want the reaction to start until we are ready, we will put some of the reagents into one flask and the rest into another, selecting them so that no reaction occurs until the contents of the two flasks are mixed. Using a 10-ml graduated cylinder to measure volumes, measure out 10 ml 0.010 M KI, 10 ml 0.001 M $Na_2S_2O_3$, and 10 ml distilled water into a 250-ml Erlenmeyer flask (Reaction Flask I). Then measure out 10 ml 0.040 M $KBrO_3$ and 10 ml 0.10 M HCl into a 125-ml Erlenmeyer flask (Reaction Flask II). To Flask II add 3 or 4 drops of starch indicator solution.

Pour the contents of Reaction Flask II into Reaction Flask I and swirl the solutions to mix them thoroughly. Note the time at which the solutions were mixed. Continue swirling the solution. It should turn blue in less than 2 minutes. Note the time at the instant that the blue color appears. Record the temperature of the blue solution to 0.2°C.

Repeat the procedure with the other mixtures in Table 21.1. *Don't forget to add the indicator* before mixing the solutions in the two flasks. The reaction flasks should be rinsed with distilled water between runs. When measuring out reagents, rinse the graduated cylinder with distilled water after you have added the reagents to Reaction Flask I, and before you measure out the reagents for Reaction Flask II. Try to keep the temperature just about the same in all the runs. Repeat any experiments that did not appear to proceed properly.

B.   Dependence of Reaction Rate on Temperature.   In this part of the experiment, the reaction will be carried out at several different temperatures, using Reaction Mixture 1 in all cases. The temperatures we will use will be about 20°C, 40°C, 10°C, and 0°C.

We will take the time at about 20°C to be that for Reaction Mixture 1 as determined at room temperature. To determine the time at 40°C proceed as follows. Make up Reaction Mixture 1 as you did in Part A, including the indicator. However, instead of mixing the solutions in the two flasks at room temperature, put the flasks into water at 40°C, drawn from the hot water tap into one or more large beakers. Check to see that the water is indeed at about 40°C, and leave the flasks in the water for several minutes to bring them to the proper temperature. Then mix the two solutions, noting the time of mixing. Continue swirling the reaction flask in the warm water. When the color change occurs note the time and the temperature of the solution in the flask.

Repeat the experiment at about 10°C, cooling all the reactants in water at that temperature before starting the reaction. Record the time required for the color to change and the final temperature of the reaction mixture. Repeat once again at about 0°C, this time using an ice-water bath to cool the reactants.

C.   Dependence of the Reaction Rate on the Presence of Catalyst (Optional).   Some ions have a pronounced catalytic effect on the rates of many reactions in water solution. Observe the effect on this reaction by once again making up Reaction Mixture 1. Before mixing, add 1 drop 0.5 M $(NH_4)_2MoO_4$, ammonium molybdate, and a few drops of starch indicator to Reaction Flask II. Swirl the flask to mix the catalyst thoroughly. Then mix the solutions, noting the time required for the color to change.

DATA AND CALCULATIONS: Rates of Chemical Reactions, II. A Clock Reaction

A. Orders of the Reaction. Rate Constant Determination

Reaction: $6\,I^-(aq) + BrO_3^-(aq) + 6\,H^+(aq) \rightarrow 3\,I_2(aq) + Br^-(aq) + 3\,H_2O$ (1)

$$\text{rate} = k\,(I^-)^m(BrO_3^-)^n(H^+)^p = -\frac{\Delta(BrO_3^-)}{t}$$ (2)

In all the reaction mixtures used in the experiment, the color change occurred when a constant predetermined number of moles of BrO_3^- had been used up by the reaction. The color "clock" allows you to measure the *time required* for this *fixed number of moles* of BrO_3^- to react. The rate of each reaction is determined by the time t required for the color to change; since in Equation 2 the change in concentration of BrO_3^- ion, $\Delta(BrO_3^-)$, is the same in each mixture, the relative rate of each reaction is inversely proportional to the time t. Since we are mainly concerned with relative rather than absolute rate, we will for convenience take all relative rates as being equal to $1000/t$. Fill in the table below, first calculating the relative reaction rate for each mixture.

Reaction mixture	Time t(sec) for color to change	Relative rate of reaction, $1000/t$	Reactant concentrations in reacting mixture (M)			Temp. in °C
			(I^-)	(BrO_3^-)	(H^+)	
1	————	————	0.0020	————	————	————
2	————	————	————	————	————	————
3	————	————	————	————	————	————
4	————	————	————	————	————	————
5	————	————	————	————	————	————

The reactant concentrations in the reaction mixture are *not* those of the stock solutions, since the reagents were diluted by the other solutions. The final volume of the reaction mixture is 50 ml in all cases. Since the number of moles of reactant does not change on dilution we can say, for example, for I^- ion, that

$$\text{no. moles } I^- = (I^-)_{stock} \times V_{stock} = (I^-)_{mixture} \times V_{mixture}$$

For Reaction Mixture 1, $(I^-)_{stock} = 0.010\text{ M}, V_{stock} = 10\text{ ml}, V_{mixture} = 50\text{ ml}$

Therefore, $(I^-)_{mixture} = \dfrac{0.010\text{ M} \times 10\text{ ml}}{50\text{ ml}} = 0.0020\text{ M}$

Calculate the rest of the concentrations in the table by the same approach.

Continued on following page   **175**

Determination of the Orders of the Reaction

Given the data in the table, the problem is to find the order for each reactant and the rate constant for the reaction. Since we are dealing with relative rates, we can modify Equation 2 to read as follows:

$$\text{relative rate} = k'(I^-)^m(BrO_3^-)^n(H^+)^p \tag{5}$$

We need to determine the relative rate constant k' and the orders m, n, and p in such a way as to be consistent with the data in the table.

The solution to this problem is quite simple, once you make a few observations on the reaction mixtures. Each mixture (2 to 4) differs from Reaction Mixture 1 in the concentration of only one species (see table). This means that for any pair of mixtures that includes Reaction Mixture 1, there is only one concentration that changes. From the ratio of the relative rates for such a pair of mixtures we can find the order for the reactant whose concentration was changed. Proceed as follows:

Write Equation 5 below for Reaction Mixtures 1 and 2, substituting the relative rates and the concentrations of I^-, BrO_3^-, and H^+ ions from the table you have just completed.

Relative rate 1 = _____ = $k'($ $)^m ($ $)^n ($ $)^p$

Relative rate 2 = _____ = $k'($ $)^m ($ $)^n ($ $)^p$

Divide the first equation by the second, noting that nearly all the terms cancel out! The result is simply

$$\frac{\text{Relative rate 1}}{\text{Relative rate 2}} =$$

If you have done this properly, you will have an equation involving only m as an unknown. Solve this equation for m, the order of the reaction with respect to I^- ion.

$$m = \underline{\qquad} \text{ (nearest integer)}$$

Applying the same approach to Reaction mixtures 1 and 3, find the value of n, the order of the reaction with respect to BrO_3^- ion.

Relative rate 1 = _____ = $k'($ $)^m ($ $)^n ($ $)^p$

Relative rate 3 = _____ = $k'($ $)^m ($ $)^n ($ $)^p$

Dividing one equation by the other:

$$=$$

$$n = \underline{\qquad}$$

Continued on following page

Now that you have the idea, apply the method once again, this time to Reaction Mixtures 1 and 4, and find p, the order with respect to H^+ ion.

Relative rate $4 = k'($ $)^m ($ $)^n ($ $)^p$

Dividing the equation for Relative rate 1 by that for Relative rate 4, we get

$$=$$

$$p = \underline{\hspace{2cm}}$$

Having found m, n, and p (nearest integers), the relative rate constant, k', can be calculated by substitution of m, n, p, and the known rates and reactant concentrations into Equation 5. Evaluate k' for Reaction Mixtures 1 to 4.

Reaction	1	2	3	4	
k'	_____	_____	_____	_____	k'_{ave} _____

Why should k' have nearly the same value for each of the above Reactions?

Using k'_{ave} in Equation 5, predict the relative rate and time t for Reaction Mixture 5. Use the concentrations in the table.

relative rate$_{pred}$ _____ t_{pred} _____ t_{obs} _____

B.   The Effect of Temperature on Reaction Rate: The Activation Energy.   To find the activation energy for the reaction it will be helpful to complete the table below.

The dependence of the rate constant, k', for a reaction is given by Equation 4:

$$\log_{10} k' = \frac{-E_a}{2.3RT} + \text{constant} \tag{4}$$

Since the reactions at the different temperatures all involve the same reactant concentrations, the rate constants, k', for two different mixtures will have the same ratio as the reaction rates themselves for the two mixtures. This means that in the calculation of E_a, we can use the observed relative rates instead of rate constants. Proceeding as before, calculate the relative rates of reaction in each of the mixtures and enter these values in (c) below. Take the $\log_{10}$ rate for each mixture and enter these values in (d). To set up the terms in $1/T$, fill in (b), (e), and (f) in the table.

Continued on following page

<u>Approximate temperature in °C</u>

	20	40	10	0
(a) Time t in seconds for color to appear	_____	_____	_____	_____
(b) Temperature of the reaction mixture in °C	_____	_____	_____	_____
(c) Relative rate = $1000/t$	_____	_____	_____	_____
(d) Log_{10} of relative rate	_____	_____	_____	_____
(e) Temperature T in K	_____	_____	_____	_____
(f) $1/T$, K^{-1}	_____	_____	_____	_____

To evaluate E_a, make a graph of log relative rate vs. $1/T$ on the graph paper provided. Find the slope of the line obtained by drawing the best straight line through the experimental points. Note that each square on the abscissa equals 0.00005 and each square on the ordinate equals 0.2.

Slope = _____

The slope of the line equals $-E_a/2.3\ R$, where $R = 8.31$ joules/mole K if E_a is to be in joules per mole. Find the activation energy, E_a, for the reaction.

E_a = _____ joules

C.   (Optional) Effect of a Catalyst on Reaction Rate

	Reaction 1	Catalyzed Reaction 1
Time for color to appear (seconds)	_____	_____

Would you expect the activation energy, E_a, for the catalyzed reaction to be greater than, less than, or equal to the activation energy for the uncatalyzed reation? Why?

RATES OF CHEMICAL REACTIONS, II. A CLOCK REACTION
(Data and Calculations)

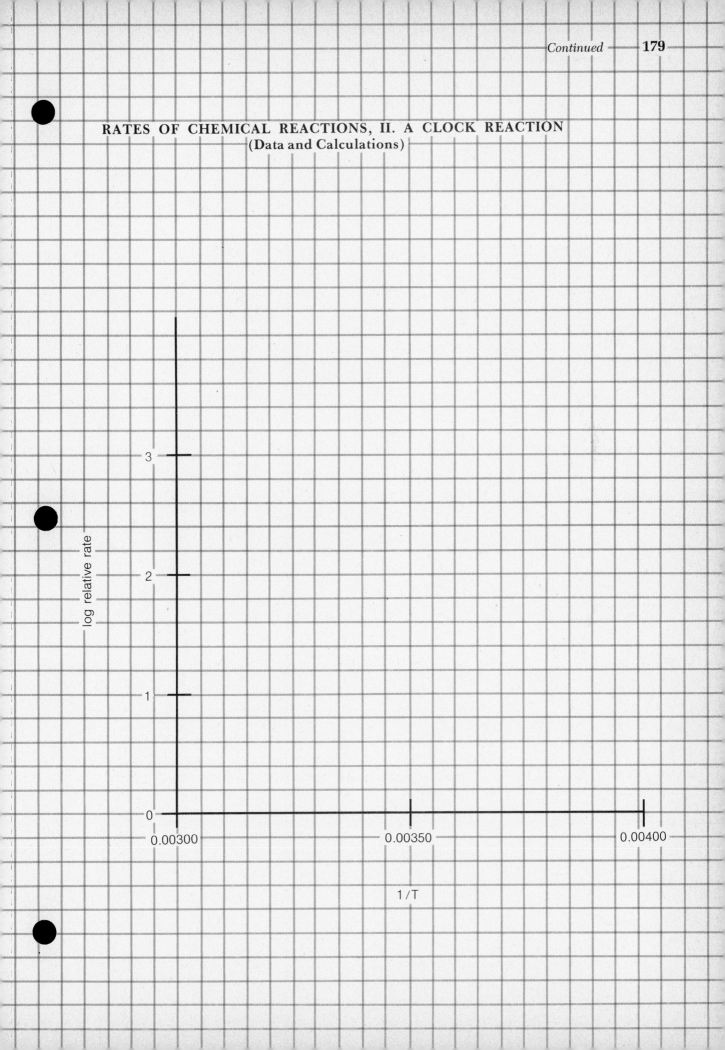

ADVANCE STUDY ASSIGNMENT: Rate of Reactions, II.
                                             A Clock Reaction

1.   A student studied the clock reaction described in this experiment. She set up Reaction Mixture 2 by mixing 20 ml 0.010 M KI, 10 ml 0.001 M $Na_2S_2O_3$, 10 ml 0.040 M $KBrO_3$, and 10 ml 0.10 M HCl using the procedure given. It took about 45 seconds for the color to turn blue.

    a. She found the concentrations of each reactant in the reacting mixture by realizing that the number of moles of each reactant did not change when that reactant was mixed with the others, but that its concentration did. For any reactant A,

$$\text{no. moles A} = M_{\text{A stock}} \times V_{\text{stock}} = M_{\text{A mixture}} \times V_{\text{mixture}}$$

The volume of the mixture was 50 ml. Revising the above equation, she obtained

$$M_{\text{A mixture}} = M_{\text{A stock}} \times \frac{V_{\text{stock}}(\text{ml})}{50 \text{ ml}}$$

Find the concentrations of each reactant by using the above equation.

$$(\text{I}^-) = \underline{\hspace{2cm}} \text{M}; \; (\text{BrO}_3{}^-) = \underline{\hspace{2cm}} \text{M}; \; (\text{H}^+) = \underline{\hspace{2cm}} \text{M}$$

    b. What was the relative rate of the reaction $(1000/t)$? _____

    c. Knowing the relative rate of reaction for Mixture 2 and the concentrations of I^-, $\text{BrO}_3{}^-$, and H^+ in that mixture, she was able to set up Equation 5 for the relative rate of the reaction. The only quantities that remained unknown were k', m, n, and p. Set up Equation 5 as she did, presuming she did it properly. Equation 5 is on page 176.

2.   For Reaction Mixture 1 the student found that 85 seconds were required. On dividing Equation 5 for Reaction Mixture 1 by Equation 5 for Reaction Mixture 2, and after canceling out the common terms (k', terms in $(\text{BrO}_3{}^-)$ and (H^+)), she got the following equation:

$$\frac{11.8}{22} = \left(\frac{0.0020}{0.0040}\right)^m = \left(\frac{1}{2}\right)^m$$

Recognizing that 11.8/22 is about equal to $\frac{1}{2}$, she obtained an approximate value for m. What was that value?

$$m = \underline{\hspace{2cm}}$$

Continued on following page   **181**

By taking logarithms of both sides of the equation, she got an exact value for m. What was that value?

$$m = \underline{\hspace{2cm}}$$

Since orders of reactions are often integers, she reported her approximate value as the order of the reaction with respect to I^-.

EXPERIMENT

22 • Analysis of an Unknown Chloride Solution

One of the important applications of precipitation reactions lies in the area of quantitative analysis. Many substances that can be precipitated from solution are so slightly soluble that the precipitation reaction by which they are formed can be considered to proceed to completion. Silver chloride is an example of such a substance.

$$AgCl(s) \rightleftharpoons Ag^+(aq) + Cl^-(aq) \qquad K_{sp} = [Ag^+][Cl^-] = 1.6 \times 10^{-10}$$

Although silver chloride is in chemical equilibrium with its ions in solution, the equilibrium constant K_{sp} for the reaction is so low that if AgCl is precipitated by the addition of a solution containing Ag^+ ion to one containing Cl^- ion, essentially all the Ag^+ added will precipitate as AgCl until essentially all the Cl^- is used up. When the amount of Ag^+ added to the solution is equal to the amount of Cl^- originally present, the precipitation of Cl^- ion will be, for all practical purposes, complete.

A convenient method for chloride analysis using AgCl has been devised. A solution of $AgNO_3$ is added to a chloride solution just to the point where the number of moles of Ag^+ added is equal to the number of moles of Cl^- initially present. We analyze for Cl^- by simply measuring how many moles of $AgNO_3$ are required. Surprisingly enough, this measurement is rather easily made by an experimental procedure called a *titration*.

In the titration a solution of $AgNO_3$ of known concentration (in moles $AgNO_3$ per liter of solution) is added from a calibrated buret to a solution containing a measured amount of unknown. The titration is stopped when a color change occurs in the solution, indicating that stoichiometrically equivalent amounts of Ag^+ and Cl^- are present. The color change is caused by a chemical reagent, called an indicator, which is added to the solution at the beginning of the titration.

The volume of $AgNO_3$ solution that has been added up to the time of the color change can be measured accurately with the buret, and the number of moles of Ag^+ added can be calculated from the known concentration of the solution.

In the Mohr method for the volumetric analysis of chloride, which we will employ in this experiment, the indicator used is K_2CrO_4. The chromate ion present in solutions of this substance will react with silver ion to form a red precipitate of Ag_2CrO_4. This precipitate will form as soon as $[Ag^+]^2 \times [CrO_4^{2-}]$ exceeds the solubility product of Ag_2CrO_4, which is about 1×10^{-12}. Under the conditions of the titration, the Ag^+ added to the solution reacts preferentially with Cl^- until that ion is essentially quantitatively removed from the system, at which point Ag_2CrO_4 begins to precipitate and the solution color changes from yellow to buff. The end point of the titration is that point at which the color change is first observed.

In this experiment, solutions containing an unknown concentration of chloride will be titrated with a standardized solution of $AgNO_3$, and the volumes of $AgNO_3$ solution required to reach the end point of each titration will be measured. Given the molarity of the $AgNO_3$

$$\text{no. of moles } Ag^+ = \text{no. of moles } AgNO_3 = M_{AgNO_3} \times V_{AgNO_3} \qquad (1)$$

where the volume of $AgNO_3$ is expressed in liters and the molarity M_{AgNO_3} is in moles per liter of solution. At the end point of the titration,

$$\text{no. of moles } Ag^+ \text{ added} = \text{no. of moles } Cl^- \text{ present in unkown} \qquad (2)$$

The concentration of chloride ion in moles per liter is simply equal to the number of moles of Cl^- ion present in the titrated sample divided by the volume of that sample in liters:

$$[Cl^-] = \frac{\text{no. of moles } Cl^- \text{ present in titrated sample}}{\text{volume of titrated sample in liters}} \qquad (3)$$

In our titrations the volume of the titrated sample will be the same in all cases and equal to 10.0 ml, or 0.0100 liter.

If we wish to find the number of grams of chloride ion in the titrated sample we multiply the number of moles by the mass of one mole of chloride ion, which is numerically equal to the atomic mass of chlorine, 35.45.

$$\text{no. of grams } Cl^- = \text{no. of moles } Cl^- \times \frac{35.45 \text{ g}}{1 \text{ mole}} \qquad (4)$$

EXPERIMENTAL PROCEDURE

Take a clean, *dry* 250-ml Erlenmeyer flask to the stockroom. You will be given about 50 ml of an unknown chloride solution, a buret, and a 10-ml pipet.

Go to the reagent shelf and, using the graduated cylinder there, measure out about 100 ml of standardized $AgNO_3$ solution into another clean, *dry* 250-ml Erlenmeyer flask. This will be your total supply for the entire experiment, so do not waste it. Clean your buret thoroughly with soap solution and rinse it with distilled water. Pour three successive 2- or 3-ml portions of the $AgNO_3$ solution into the buret and tip it back and forth to rinse the inside walls. Allow the $AgNO_3$ to drain out of the buret tip completely each time. Fill the buret with the $AgNO_3$ solution. Open the stopcock momentarily to flush any air bubbles out of the tip of the buret. Be sure that your stopcock fits snugly and that the buret does not leak.

Pipet 10 ml of your unknown chloride solution into a clean, but not necessarily dry, 125-ml Erlenmeyer flask. This is the sample you will be titrating. To that sample add about 25 ml distilled water, to give you enough liquid volume to work with, and 3 drops of 1 M K_2CrO_4, which will give the solution a yellow color.

Read the initial buret level to 0.02 ml. You may find it useful when making readings to put a white card marked with a thick, black stripe behind the meniscus. If the black line is held just below the level to be read, its reflection in the surface of the meniscus will help you obtain a very accurate reading. Begin to add the $AgNO_3$ solution to the chloride solution in the Erlenmeyer flask. A white precipitate of AgCl will form immediately, and the amount will increase during the course of the titration. At the beginning of the titration, you can add the $AgNO_3$ fairly rapidly, a few ml at a time, swirling the flask as best you can to mix the solution. You will find that at the point where the $AgNO_3$ hits the solution, there will be a red trace of Ag_2CrO_4, which disappears when you stop adding nitrate and swirl the flask. As you proceed with the titration, the red color will persist more and more, since the amount of excess chloride ion, which reacts with the Ag_2CrO_4 to form AgCl, will slowly decrease. Gradually decrease the rate at which you add $AgNO_3$ as the red color becomes stronger. At some stage you may find it convenient to set your buret stopcock to deliver $AgNO_3$ slowly, drop by drop, while you swirl the flask. When you are near the end point, add the $AgNO_3$ drop by drop, swirling between drops. The end point of the titration is that point where the mixture first takes on a permanent reddish-yellow or buff color which does not revert to pure yellow on swirling. If you are careful, you can hit the end point within a few drops of $AgNO_3$. When you have reached the end point, stop the titration and record the buret level.

Pour the solution you have just titrated into another 125-ml Erlenmeyer flask or into a 250-ml beaker. To that solution add a few ml of your unknown chloride solution. The color of the mixture should revert to the original yellow. Use the color of this mixture as a reference against which you can compare your samples in the remaining titrations.

Using distilled water, rinse out the 125-ml Erlenmeyer flask in which you carried out the titration. Then pipet another 10-ml sample of unknown (sometimes called an aliquot) into the flask. Add about 25 ml distilled water and the K_2CrO_4 indicator. Refill your buret, take a volume reading, and titrate the solution as before. This titration should be more accurate than the first, for two reasons. The sample volume to be titrated is the same as before, so the volume of $AgNO_3$ will be nearly the same as before. In addition, you have a reference for color comparison which should make it easier to recognize when a color change has occurred.

Titrate the third sample as you did the second. With care it should be possible to obtain volumes of $AgNO_3$ which agree to within less than 1% in the last two titrations.

Silver nitrate is very expensive. Pour all titrated solutions and any $AgNO_3$ solution remaining in your buret or flask into the waste bottles provided.

DATA AND CALCULATIONS: Analysis of an Unknown Chloride

Unknown sample no. _____

Molarity of standardized $AgNO_3$ solution _____ M

	I	II	III
Initial buret reading	_____ ml	_____ ml	_____ ml
Final buret reading	_____ ml	_____ ml	_____ ml
Volume of $AgNO_3$ used in titration of 10-ml sample of unknown	_____ ml	_____ ml	_____ ml
No. of moles $AgNO_3$ used	_____ moles	_____ moles	_____ moles
No. of moles Cl^- present in 10-ml sample	_____ moles	_____ moles	_____ moles
Molarity of Cl^- in unknown	_____ M	_____ M	_____ M
Value of molarity that you wish to report	_____ M		

If you didn't select the average molarity as your reported value, state your reasons:

No. of grams of Cl^- present in   _____ g
10-ml sample of unknown

No. of grams of Cl^- present in   _____ g
1 liter of unknown

ADVANCE STUDY ASSIGNMENT: Analysis of an Unknown Chloride

1. A student pipets 10 ml of an unknown chloride solution into a 125-ml Erlenmeyer flask. He adds 25 ml water and 3 drops 1 M K_2CrO_4. He cleans his buret and fills it with 0.02268 M $AgNO_3$. The initial volume reading on the buret is 2.12 ml. He titrates the unknown solution until the color just turns from yellow to buff and finds that the volume reading on the buret is 33.59 ml.

　　a. What volume of $AgNO_3$ did he use?

_____ml = _____liters

　　b. How many moles of $AgNO_3$ were in that volume (Eq. 1)?

_____moles

　　c. How many moles of Ag^+ were in that volume of solution?　　_____moles

　　d. How many moles of Cl^- were there in the sample that was titrated (Eq. 2)?

_____moles

　　e. How many grams of Cl^- were in the titrated sample (Eq. 4)?

_____grams

　　f. What volume of unknown solution was used in the titration?

_____ml = _____liters

　　g. What was the concentration of Cl^- in the *unknown* in moles per liter (Eq. 3).

_____M

EXPERIMENT

23 • Determination of the Solubility Product of PbI$_2$

In this experiment you will determine the solubility product of lead iodide, PbI$_2$. Lead iodide is relatively insoluble, having a solubility of less than 0.002 mole per liter at 20°C. The equation for the solution reaction of PbI$_2$ is

$$PbI_2(s) \rightleftarrows Pb^{2+}(aq) + 2I^-(aq) \tag{1}$$

The solubility product expression associated with this reaction is

$$K_{sp} = [Pb^{2+}][I^-]^2 \tag{2}$$

Equation 2 implies that in any system containing solid PbI$_2$ in equilibrium with its ions, the product of $[Pb^{2+}]$ times $[I^-]^2$ will at a given temperature have a fixed magnitude, independent of how the equilibrium system was initially made up.

In the first part of the experiment, known volumes of standard solutions of Pb(NO$_3$)$_2$ and KI will be mixed in several different proportions. The yellow precipitate of PbI$_2$ formed will be allowed to come to equilibrium with the solution. The value of $[I^-]$ in the solution will be measured experimentally. The $[Pb^{2+}]$ will be calculated from the initial composition of the system, the measured value of $[I^-]$, and the stoichiometric relation between Pb^{2+} and I$^-$ in Equation 1.

By mixing the solutions as we have described, we approach equilibrium by precipitating PbI$_2$ and measuring the concentrations of I$^-$ and Pb^{2+} remaining in the solution. We will also carry out the reaction in the other direction, by first precipitating PbI$_2$, washing it free of excess ions, and then dissolving the solid in the inert salt solution. Under such conditions the concentrations of Pb^{2+} and I$^-$ in the saturated solution will be related by Equation 1, since both ions come from pure PbI$_2$. From the measured value of $[I^-]$ in the saturated solution we can calculate $[Pb^{2+}]$ immediately.

The concentration of I$^-$ ion will be found spectrophotometrically, as in Experiment 17. Although the iodide ion is not colored, it is relatively easily oxidized to I$_2$, which is brown in water solution. Our procedure will be to separate the solid PbI$_2$ from the solution and then to oxidize the I$^-$ in solution with potassium nitrite, KNO$_2$, under slightly acidic conditions, where the conversion to I$_2$ is quantitative. Although the concentration of I$_2$ will be rather low in the solutions you will prepare, the absorption of light by I$_2$ in the vicinity of 525 nm is sufficiently intense to make accurate analyses possible.

In all of the solutions prepared, potassium nitrate KNO$_3$ (note this distinction between KNO$_2$ and KNO$_3$!) will be present as an inert salt. This salt serves to keep the ionic strength of the solution essentially constant at 0.2 M and promotes the formation of well-defined crystalline precipitates of PbI$_2$.

EXPERIMENTAL PROCEDURE

From the stock solutions that are available, measure out about 35 ml of 0.012 M Pb(NO$_3$)$_2$ in 0.20 M KNO$_3$ into a small beaker. To a second small beaker add 30 ml 0.030 M KI in 0.20 M KNO$_3$ and, to a third, add 10 ml 0.20 M KNO$_3$. Use the labeled graduated cylinders next to each of the reagent bottles for measuring out these solutions. Use these reagent solutions in your experiment.

Label five regular test tubes 1 to 5, either with labels or by noting their positions in your test tube rack. Into the first four tubes pipet 5.0 ml of 0.012 M Pb(NO$_3$)$_2$ in KNO$_3$. Then, to test tube 1, add 2.0 ml 0.030 M KI in KNO$_3$. Add 3.0, 4.0, and 5.0 ml of that same solution to test tubes

191

2, 3, and 4, respectively. Add enough 0.20 M KNO$_3$ to the first three test tubes to make the total volume 10.0 ml in each tube. The composition of the final mixture in each tube is summarized in Table 23.1.

TABLE 23.1   VOLUMES OF REAGENTS USED IN PRECIPITATING PbI$_2$ (ML)

Test Tube	0.012 M Pb(NO$_3$)$_2$	0.030 M KI	0.20 M KNO$_3$
1	5.0	2.0	3.0
2	5.0	3.0	2.0
3	5.0	4.0	1.0
4	5.0	5.0	0.0
5	10.0	10.0	0.0

Stopper each test tube and shake thoroughly at intervals of several minutes while you are proceeding with the next part of the experiment.

In the fifth test tube mix about 10 ml of 0.012 M Pb(NO$_3$)$_2$ in KNO$_3$ with 10 ml of 0.03 M KI in KNO$_3$. Shake the mixture vigorously for a minute or so. Let the solid settle for a few minutes and then decant and discard three-fourths of the supernatant solution. Transfer the solid PbI$_2$ and the rest of the solution to a small test tube and centrifuge. Discard the liquid, retaining the solid precipitate. Add 3 ml 0.20 M KNO$_3$ and shake to wash the solid free of excess Pb^{2+} or I$^-$. Centrifuge again, and discard the liquid. By this procedure you should now have prepared a small sample of essentially pure PbI$_2$ in a little KNO$_3$ solution. Add 0.20 M KNO$_3$ to the solid until the tube is about three-fourths full. Shake well at several one minute intervals to saturate the solution with PbI$_2$.

In this experiment it is essential that the volumes of reagents used to make up the mixtures in test tubes 1 to 4 be measured accurately. It is also essential that all five mixtures be shaken thoroughly so that equilibrium can be established. Insufficient shaking of the first four test tubes will result in not enough PbI$_2$ precipitating to reach true equilibrium; if the small test tube is not shaken sufficiently, not enough PbI$_2$ will dissolve to attain equilibrium.

When each of the mixtures has been shaken for at least 15 minutes, let the tubes stand for three to four minutes to let the solid settle. Pour the supernatant liquid in test tube 1 into a small dry test tube until it is three-fourths full and centrifuge for about three minutes to settle the solid PbI$_2$. Pour the liquid into another small dry test tube; if there are any solid particles or yellow color remaining in the liquid, centrifuge again. When you have a clear liquid, dip a small piece of clean, dry paper towel into the liquid to remove floating PbI$_2$ particles from the surface. Pipet 3.0 ml of 0.02 M KNO$_2$, potassium NITRITE (*not* KNO$_3$, potassium nitrate), into a clean, dry spectrophotometer tube and add 2 drops 6 M HCl. Then, using a medicine dropper, add enough of the clear centrifuged solution (about 3 ml) to fill the spectrophotometer tube just to the level indicated on the tube. Shake gently to mix the reagents and then measure the absorbance of the solution as directed by your instructor. The calibration curve or equation which is provided will allow you to determine directly the concentration of I$^-$ ion that was in equilibrium with PbI$_2$. Use the same procedure to analyze the solutions in test tubes 2 through 5, completing each analysis before you proceed to the next.

DATA AND CALCULATIONS: Determination of the Solubility Product of PbI_2

From the experimental data we obtain $[I^-]$ directly. To obtain K_{sp} for PbI_2 we must calculate $[Pb^{2+}]$ in each equilibrium system. This is most easily done by constructing an equilibrium table. We first find the initial amounts of I^- and Pb^{2+} ions in each system from the way the mixtures were made up. Knowing $[I^-]$ and the formula of lead iodide allows us to calculate $[Pb^{2+}]$. K_{sp} then follows directly. The calculations are similar to those in Experiment 19. See the following page if you need help with calculations.

$$PbI_2(s) \rightleftharpoons Pb^{2+}(aq) + 2I^-(aq) \qquad K_{sp} = [Pb^{2+}][I^-]^2$$

DATA

Test tube no.	1	2	3	4	5
ml 0.012 M $Pb(NO_3)_2$	_____	_____	_____	_____	Saturated soln. of PbI_2
ml 0.03 M KI	_____	_____	_____	_____	_____
ml 0.20 M KNO_3	_____	_____	_____	_____	_____
Total volume in ml	_____	_____	_____	_____	_____
Absorbance of solution	_____	_____	_____	_____	_____
$[I^-]$ in moles/liter at equilibrium	_____	_____	_____	_____	_____

CALCULATIONS

1. Initial no. moles Pb^{2+} ____ $\times 10^{-5}$ ____ $\times 10^{-5}$ ____ $\times 10^{-5}$ ____ $\times 10^{-5}$

2. Initial no. moles I^- ____ $\times 10^{-5}$ ____ $\times 10^{-5}$ ____ $\times 10^{-5}$ ____ $\times 10^{-5}$

3. No. moles I^- at equilibrium   ____ $\times 10^{-5}$ ____ $\times 10^{-5}$ ____ $\times 10^{-5}$ ____ $\times 10^{-5}$

4. No. moles I^- precipitated   ____ $\times 10^{-5}$ ____ $\times 10^{-5}$ ____ $\times 10^{-5}$ ____ $\times 10^{-5}$

5. No. moles Pb^{2+} precipitated   ____ $\times 10^{-5}$ ____ $\times 10^{-5}$ ____ $\times 10^{-5}$ ____ $\times 10^{-5}$

6. No. moles Pb^{2+} at equilibrium   ____ $\times 10^{-5}$ ____ $\times 10^{-5}$ ____ $\times 10^{-5}$ ____ $\times 10^{-5}$

7. $[Pb^{2+}]$ at equilibrium   _____ _____ _____ _____ _____

8. K_{sp} PbI_2 _____ _____ _____ _____ _____

Suggestions Regarding Calculations

In test tubes 1 to 4, calculations must be based on the amounts of reagents which were used. In test tube 5, Sections 1–6 do not apply; the solution is saturated with PbI_2, so initial amounts or reagents are not relevant. Several of the calculations depend upon the definition of molarity. For species A,

$$M_A = \frac{\text{no. moles of A}}{\text{volume of soln}} \qquad \text{or no. moles of A} = M_A \times \text{volume (liters)} \qquad (6)$$

1. The initial number of moles of Pb^{2+} is equal to the amount in the $Pb(NO_3)_2$ solution that was used. What is $[Pb^{2+}]$ in that solution? _____M. What volume was used? _____ ml; _____ liters. Use Equation 6 to find the initial number of moles of Pb^{2+}. Since the volumes used were equal, each tube (1–4) contains the same number of moles of Pb^{2+}.

2. What is $[I^-]$ in the KI solution? _____M. What volumes were used? _____ liters. Use Equation 6 to find the initial numbers of moles of I^- present. Since the solution volumes in test tubes 1 to 4 differ, the numbers of moles of I^- differ.

3. Here we need to find the number of moles of I^- in each solution at equilibrium. What is the total volume of each solution? _____ml; _____liters (same for tubes 1 to 4). From $[I^-]$ as measured at equilibrium, and the solution volume, find the number of moles of I^- in each solution at equilibrium, again using Equation 6.

4. The number of moles of I^- in solution at equilibrium is given by the equation.

$$\text{no. moles } I^- \text{ in soln} = \text{initial no. moles } I^- - \text{no. moles } I^- \text{ precipitated}$$

From the no. of moles of I^- you started with (Section 2) and the number remaining in solution (Section 3), you can easily obtain the number of moles of I^- that precipitated.

5. The number of moles of Pb^{2+} that precipitated is related to the number of moles of I^- that precipitated, since they both come down in the PbI_2. What is the relationship? _____ _____ Calculate the number of moles Pb^{2+} that precipitated from the number of moles of I^- that precipitated.

6. The number of moles Pb^{2+} in solution at equilibrium must equal the number initially present minus the number that precipitated. Make that calculation for tubes 1 to 4.

7. What was the total volume of solution for each of the test tubes 1 to 4? _____ ml; _____ liters. Use Equation 6 to find $[Pb^{2+}]$ in the solution in each test tube. In the solution in test tube 5, the Pb^{2+} and the I^- both came from dissolving PbI_2, so the concentrations of those two ions are related. What is the relationship? _____ Calculate $[Pb^{2+}]$ from $[I^-]$ in tube 5.

8. From $[Pb^{2+}]$ and $[I^-]$ at equilibrium, find K_{sp} for PbI_2 for each of the solutions in test tubes 1 to 5, using Equation 2.

Name _____ **Section** _____

ADVANCE STUDY ASSIGNMENT: Determination of the Solubility Product of PbI_2

1. State in words the meaning of the solubility product equation for PbI_2:

$$K_{sp} = [Pb^{2+}][I^-]^2$$

2. When 5.0 ml of 0.012 M $Pb(NO_3)_2$ are mixed with 5.0 ml of 0.030 M KI, a yellow precipitate of $PbI_2(s)$ forms.

   a. How many moles of Pb^{2+} are initially present? (See Section 1, Suggestions Regarding Calculations, if you wish to have help.)

   _____ moles

   b. How many moles of I^- are originally present (Section 2)?

   _____ moles

   c. In a colorimeter the equilibrium solution is analyzed for I^-, and its concentration is found to be 7×10^{-3} mole/liter. How many moles of I^- are present in the solution (10 ml) (Section 3)?

   _____ moles

   d. How many moles of I^- precipitated (Section 4)?

   _____ moles

   e. How many moles of Pb^{2+} precipitated (Section 5)?

   _____ moles

   f. How many moles of Pb^{2+} are left in solution (Section 6)?

   _____ moles

   g. What is the concentration of Pb^{2+} in the equilibrium solution (Section 7)?

   _____ moles/liter

   h. Find a value for K_{sp} of PbI_2 from these data (Section 8).

   _____

EXPERIMENT

24 • The Standardization of a Basic Solution and the Determination of the Equivalent Mass of a Solid Acid

When a solution of a strong acid is mixed with a solution of a strong base, a chemical reaction occurs that can be represented by the following net ionic equation:

$$H^+(aq) + OH^-(aq) \rightarrow H_2O$$

This is called a neutralization reaction, and chemists use it extensively to change the acidic or basic properties of solutions. The equilibrium constant for the reaction is about 10^{14} at room temperature, so that the reaction can be considered to proceed completely to the right, using up whichever of the ions is present in the lesser amount and leaving the solution either acidic or basic, depending on whether H^+ or OH^- ion was in excess.

Since the reaction is essentially quantitative, it can be used to determine the concentrations of acidic or basic solutions. A frequently used procedure involves the titration of an acid with a base. In the titration, a basic solution is added from a buret to a measured volume of acid solution until the number of moles of OH^- ion added is just equal to the number of moles of H^+ ion present in the acid. At that point the volume of basic solution that has been added is measured.

Recalling the definition of the molarity, M_A, of species A, we have

$$M_A = \frac{\text{no. moles of A}}{\text{no. liters of solution, } V} \quad \text{or no. moles of A} = M_A \times V \tag{1}$$

At the end point of the titration,

$$\text{no. moles } H^+ \text{ originally present} = \text{no. moles } OH^- \text{ added} \tag{2}$$

So, by Equation 1,

$$M_{H^+} \times V_{acid} = M_{OH^-} \times V_{base} \tag{3}$$

Therefore, if the molarity of either the H^+ or the OH^- ion in its solution is known, the molarity of the other ion can be found from the titration.

The equivalence point or end point in the titration is determined by using a chemical, called an indicator, that changes color at the proper point. The indicators used in acid-base titrations are weak organic acids or bases that change color when they are neutralized. One of the most common indicators is phenolphthalein, which is colorless in acid solutions but becomes red when the pH of the solution becomes 9 or higher.

When a solution of a strong acid is titrated with a solution of a strong base, the pH at the end point will be about 7. At the end point a drop of acid or base added to the solution will change its pH by several pH units, so that phenolphthalein can be used as an indicator in such titrations. If a weak acid is titrated with a strong base, the pH at the equivalence point is somewhat higher than 7, perhaps 8 or 9, and phenolphthalein is still a very satisfactory indicator. If, however, a solution of a weak base such as ammonia is titrated with a strong acid, the pH will be a unit or two less than 7 at the end point, and phenolphthalein will not be as good an indicator for that titration as, for example, methyl red, whose color changes from red to yellow as the pH changes from about 4 to 6. Ordinarily, indicators will be chosen so that their color change occurs at about the pH at the equivalence point of a given acid-base titration.

In this experiment you will determine the molarity of OH^- ion in an NaOH solution by titrating that solution against a standardized solution of HCl. Since in these solutions one mole of acid in solution furnishes one mole of H^+ ion and one mole of base produces one mole of OH^- ion, $M_{HCl} = M_{H^+}$ in the acid solution, and $M_{NaOH} = M_{OH^-}$ in the basic solution. Therefore the titration will allow you to find M_{NaOH} as well as M_{OH^-}.

In the second part of this experiment you will use your standardized NaOH solution to titrate a sample of a pure solid organic acid. By titrating a weighed sample of the unknown acid with your standardized NaOH solution you can, by Equation 2, find the number of moles of H^+ ion that your sample can furnish. From the mass of your sample and the number of moles of H^+ it contains, you can calculate the number of grams of solid acid, EM, that would contain one mole of H^+ ion.

$$EM = \frac{\text{no. grams acid}}{\text{no. moles } H^+ \text{ furnished}} \tag{4}$$

The value of EM for an acid is called the *equivalent mass* of the acid and is equal to the number of grams of acid per mole of available H^+ ion:

$$\text{one equivalent mass acid} \rightarrow \text{one mole } H^+ \text{ ion} \tag{5}$$

The equivalent mass, EM, of an acid may or may not equal the molar mass of the acid, MM. The reason is almost, but not quite, obvious. Let us consider three acids, with the molecular formulas HX, H_2Y, and H_3Z:

one mole HX $\rightarrow$ one mole H^+ Therefore, by Equation 5, MM = EM

one mole H_2Y $\rightarrow$ two moles H^+ "   MM = 2 × EM

one mole H_3Z $\rightarrow$ three moles H^+ "   MM = 3 × EM

In all cases the molar mass and the equivalent mass are related by simple equations, but in order to find the molar mass from the equivalent mass we need to know the molecular formula of the acid.

EXPERIMENTAL PROCEDURE

Note: This experiment is relatively long unless you know precisely what you are to do. Study the experiment carefully before coming to class, so that you don't have to spend a lot of time finding out what the experiment is all about.

Obtain two burets and a sample of solid unknown acid from the stockroom.

A. Standardization of NaOH Solution. Into a small graduated cylinder draw about 7 ml of the stock 6 M NaOH solution provided in the laboratory and dilute to about 400 ml with distilled water in a 500 ml Florence flask. Stopper the flask tightly and mix the solution thoroughly at intervals over a period of at least 15 minutes before using the solution.

Draw into a clean *dry* 125 ml Erlenmeyer flask about 75 ml of standardized HCl solution (about 0.1 M) from the stock solution on the reagent shelf. This amount should provide all the standard acid you will need; do not waste it. Record the molarity of the HCl.

Prepare for the titration by using the procedure described in Experiment 22. The purpose of this procedure is to make sure that the solution in each buret has the same molarity as it has in the container from which it was poured. Clean the two burets and rinse them with distilled water. Then rinse one buret three times with a few ml of the HCl solution, in each case thoroughly wetting the walls of the buret with the solution and then letting it out through the stopcock. Fill the buret with HCl; open the stopcock momentarily to fill the tip. Proceed to clean and fill the other buret with your NaOH solution in a similar manner. Put the acid buret, A, on the left side of your buret clamp, and the base buret, B, on the right side. Check to see that your burets do not leak and that there are no air bubbles in either buret tip. Read and record the levels in the two burets to 0.02 ml.

Draw about 25 ml of the HCl solution from the buret into a clean 250 ml Erlenmeyer flask; add to the flask about 25 ml of distilled H_2O and 2 or 3 drops of phenolphthalein indicator solution. Place a white sheet of paper under the flask to aid in the detection of any color change. Add the NaOH solution intermittently from its buret to the solution in the flask, noting the pink phenolphthalein color that appears and disappears as the drops hit the liquid and are mixed with it. Swirl the liquid in the flask gently and continuously as you add the NaOH solution. When the pink color begins to persist, slow down the rate of addition of NaOH. In the final stages of the titration add the NaOH drop by drop until the entire solution just turns a pale pink color that will persist for about 30 seconds. If you go past the end point and obtain a red solution, add a few drops of the HCl solution to remove the color, and then add NaOH a drop at a time until the pink color persists. Carefully record the final readings on the HCl and NaOH burets.

To the 250 ml Erlenmeyer flask containing the titrated solution, add about 10 ml more of the standard HCl solution. Titrate this as before with the NaOH to an end point, and carefully record both buret readings once again. To this solution add about 10 ml more HCl and titrate a third time with NaOH.

You have now completed three titrations, with *total* HCl volumes of about 25, 35, and 45 ml. Using Equation 3, calculate the molarity of your base, M_{OH^-}, for each of the three titrations. In each case, use the *total volumes* of acid and base which were added up to that point. At least two of these molarities should agree to within 1%. If they do, proceed to the next part of the experiment. If they do not, repeat these titrations until two calculated molarities do agree.

B.   Determination of the Equivalent Mass of an Acid.   Weigh the vial containing your solid acid on the analytical balance to ± 0.0001 g. Carefully pour out about half the sample into a clean but not necessarily dry 250 ml Erlenmeyer flask. Weigh the vial accurately. Add about 50 ml of distilled water and 2 or 3 drops of phenolphthalein to the flask. The acid may be relatively insoluble, so don't worry if it doesn't all dissolve.

Fill your NaOH buret with the (now standardized) NaOH solution. Add the standard HCl to your HCl buret until it is about half-full. Read both levels carefully and record them.

Titrate the solution of the solid acid with NaOH. As the acid is neutralized it will tend to dissolve in the solution. If the acid appears to be relatively insoluble, add NaOH until the pink color persists, and then swirl to dissolve the solid. If the solid still will not dissolve, and the solution remains pink, add 25 ml of ethanol to increase the solubility. If you go past the end-point, add HCl as necessary. The final pink end-point should appear on addition of one drop of NaOH. Record the final levels in the NaOH and HCl burets.

Pour the rest of your acid sample into a clean 250-ml Erlenmeyer flask, and weigh the vial accurately. Titrate this sample of acid as before with the NaOH and HCl solutions.

If you use HCl in these titrations, and you probably will, the calculations needed are a bit more complicated than in the standardization of the NaOH solution. In order to find the number of moles of H^+ ion in the solid acid, you must subtract the number of moles of HCl used from the number of moles of NaOH. For a back-titration, which is what we have in this case:

no. moles H^+ in solid acid = no. moles OH^- in NaOH soln. − no. moles H^+ in HCl soln.   (5)

For volumes in milliliters, this equation takes the form:

$$\text{no. moles } H^+ \text{ in solid acid} = \frac{M_{NaOH} \times V_{NaOH}}{1000} - \frac{M_{HCl} \times V_{HCl}}{1000} \qquad (6)$$

DATA: Standardization of a Basic Solution.
Determination of the GEM of an Acid

A. Standardization of NaOH Solution

	Trial 1	Trial 2	Trial 3
Initial reading, HCl buret	_____ ml		
Final reading, HCl buret	_____ ml	_____ ml	_____ ml
Initial reading, NaOH buret	_____ ml		
Final reading, NaOH buret	_____ ml	_____ ml	_____ ml

B. Equivalent Mass of Unknown Acid

Mass of vial plus contents	_____ g
Mass of vial plus contents less Sample 1	_____ g
Mass less Sample 2	_____ g

	Trial 1	Trial 2
Initial reading, NaOH buret	_____ ml	_____ ml
Final reading, NaOH buret	_____ ml	_____ ml
Initial reading, HCl buret	_____ ml	_____ ml
Final reading, HCl buret	_____ ml	_____ ml

CALCULATIONS

A. Standardization of NaOH Solution

	Trial 1	Trial 2	Trial 3
Total volume HCl	_____ ml	_____ ml	_____ ml
Total volume NaOH	_____ ml	_____ ml	_____ ml

Continued on following page   **201**

Molarity M_{HCl}; of standardized HCl _____ M

Molarity, M_{H^+}, in standardized HCl _____ M

By Equation 3,

$$M_{H^+} \times V_{acid} = M_{OH^-} \times V_{base} \text{ or } M_{OH^-} = M_{H^+} \times \frac{V_{HCl}}{V_{NaOH}} \tag{3}$$

Use Equation 3 to find the molarity, M_{OH^-}, of the NaOH solution. Note that the volumes do not need to be converted to liters, since we use the volume ratio.

	Trial 1	Trial 2	Trial 3

M_{OH^-} _____ M   _____ M   _____ M   (should agree with 1%)

The molarity of the NaOH will equal M_{OH^-}, since one mole NaOH $\rightarrow$ one mole OH^-.

Average molarity of NaOH solution, M_{NaOH} _____ M

B.   Equivalent Mass of Unknown Acid

	Trial 1	Trial 2
Mass of sample	_____ g	_____ g
Volume NaOH used	_____ ml	_____ ml
No. moles NaOH $= \dfrac{V_{NaOH} \times M_{NaOH}}{1000}$	_____	_____
Volume HCl used	_____ ml	_____ ml
No. moles HCl $= \dfrac{V_{HCl} \times M_{HCl}}{1000}$	_____	_____
No. moles H^+ in sample (by Eq. 5)	_____	_____
EM $= \dfrac{\text{no. grams acid}}{\text{no. moles } H^+}$	_____ g	_____ g
Unknown no.		_____

ADVANCE STUDY ASSIGNMENT: Equivalent Mass of an Unknown Acid

1. 7.0 ml of 6.0 M NaOH are diluted with water to a volume of 400 ml. We are asked to find the molarity of the resulting solution.
   a. First find out how many moles of NaOH there are in 7.0 ml of 6.0 M NaOH. Use Equation 1. Note that the volume must be in liters.

   _____moles

   b. Since the total number of moles of NaOH is not changed on dilution, the molarity after dilution can also be found by Equation 1, using the final volume of the solution. Calculate that molarity.

   _____M

2. In an acid-base titration, 24.88 ml of an NaOH solution are needed to neutralize 26.43 ml of a 0.1049 M HCl solution. To find the molarity of the NaOH solution, we can use the following procedure:

   a. First note the value of M_{H+} in the HCl solution.

   _____M

   b. Find M_{OH-} in the NaOH solution. (Use Eq. 3.)

   _____M

   c. Obtain M_{NaOH} from M_{OH-}.   _____M

3. A 0.2349-g sample of an unknown acid requires 33.66 ml of 0.1086 M NaOH for neutralization to a phenolphthalein end point. There are 0.42 ml of 0.1049 M HCl used for back-titration.

   a. How many moles of OH^- are used? How many moles of H^+ from HCl?

   _____moles OH^- _____moles H^+

   b. How many moles of H^+ are there in the solid acid? (Use Eq. 5.)

   _____moles H^+ in solid

   c. What is the equivalent mass of the unknown acid? (Use Eq. 4.)

   _____g   **203**

EXPERIMENT

25 • pH, Its Measurement and Applications

One of the more important properties of an aqueous solution is its concentration of hydrogen ion. The H^+ or H_3O^+ ion has great effect on the solubility of many inorganic and organic species, on the nature of complex metallic cations found in solutions, and on the rates of many chemical reactions. It is important that we know how to measure the concentration of hydrogen ion and understand its effect on solution properties.

For convenience the concentration of H^+ ion is frequently expressed as the pH of the solution rather than as molarity. The pH of a solution is defined by the following equation:

$$pH = -\log [H^+] \tag{1}$$

where the logarithm is taken to the base 10. If $[H^+]$ is 1×10^{-4} moles per liter, the pH of the solution is, by the equation, 4. If the $[H^+]$ is 5×10^{-2} M, the pH is 1.3.

Basic solutions can also be described in terms of pH. In water solutions the following equilibrium relation will always be obeyed:

$$[H^+] \times [OH^-] = K_w = 1 \times 10^{-14} \text{ at } 25°C \tag{2}$$

In distilled water $[H^+]$ equals $[OH^-]$, so, by Equation 2, $[H^+]$ must be 1×10^{-7} M. Therefore, the pH of distilled water is 7. Solutions in which $[H^+] > [OH^-]$ are said to be acidic and will have a pH < 7; if $[H^+] < [OH^-]$, the solution is basic and its pH > 7. A solution with a pH of 10 will have a $[H^+]$ of 1×10^{-10} M and a $[OH^-]$ of 1×10^{-4} M.

We measure the pH of a solution experimentally in two ways. In the first of these we use a chemical called an indicator, which is sensitive to pH. These substances have colors that change over a relatively short pH range (about 2 pH units) and can, when properly chosen, be used to roughly determine the pH of a solution. Two very common indicators are litmus, usually used on paper, and phenolphthalein, the most common indicator in acid-base titrations. Litmus changes from red to blue as the pH of a solution goes from about 6 to about 8. Phenolphthalein changes from colorless to red as the pH goes from 8 to 10. A given indicator is useful for determining pH only in the region in which it changes color. Indicators are available for measurement of pH in all the important ranges of acidity and basicity. By matching the color of a suitable indicator in a solution of known pH with that in an unknown solution, one can determine the pH of the unknown to within about 0.3 pH units.

The other method for finding pH is with a device called a pH meter. In this device two electrodes, one of which is sensitive to $[H^+]$, are immersed in a solution. The potential between the two electrodes is related to the pH. The pH meter is designed so that the scale will directly furnish the pH of the solution. A pH meter gives much more precise measurement of pH than does a typical indicator, and is ordinarily used when an accurate determination of pH is needed.

Some acids and bases undergo substantial ionization in water, and are called *strong* because of their essentially complete ionization in reasonably dilute solutions. Other acids and bases, because of incomplete ionization (often only about 1 per cent in 0.1 M solution), are called *weak*. Hydrochloric acid, HCl, and sodium hydroxide, NaOH are typical examples of a strong acid and a strong base respectively. Acetic acid, $HC_2H_3O_2$, and ammonia, NH_3, are classic examples of a weak acid and a weak base.

A weak acid will ionize according to the Law of Chemical Equilibrium:

$$HA(aq) \rightleftharpoons H^+(aq) + A^-(aq) \tag{3}$$

At equilibrium,

$$\frac{[H^+][A^-]}{[HA]} = K_a \tag{4}$$

K_a is a constant characteristic of the acid HA; in solutions containing HA, the product of concentrations in the equation will remain constant at equilibrium independent of the manner in which the solution was made up. A similar relation can be written for solutions of a weak base.

The value of the ionization constant K_a for a weak acid can be found experimentally in several ways. One very simple procedure involves very little calculation, is accurate, and does not even require a knowledge of the molarity of the acid. A sample of a weak acid, HA, often a solid, is dissolved in water. The solution is divided into two equal parts. One part of the solution is titrated to a phenolphthalein end point with an NaOH solution, with the HA converted to A^- by the reaction:

$$OH^-(aq) + HA(aq) \rightarrow A^-(aq) + H_2O \tag{5}$$

The number of moles of A^- produced equals the number of moles HA in the other part of the sample. The two solutions are then mixed, and the pH of the resultant solution is obtained. In that solution it is clear that [HA] equals [A^-], so that by (4),

$$[H^+] = K_a \tag{6}$$

By this method the [H^+] obtained from the pH measurement is equal to the value of the ionization constant of the acid.

Salts that can be formed by the reaction of strong acids and bases, such as NaCl, KBr, or $NaNO_3$, ionize completely but do not react with water when in solution. They form neutral solutions with a pH of about 7. When dissolved in water, salts of weak acids or weak bases furnish ions that tend to react to some extent with water, producing molecules of the weak acid or base and liberating some OH^- or H^+ ion to the solution.

If HA is a weak acid, the A^- ion produced when NaA is dissolved in water will react with water to some extent, according to the equation

$$A^-(aq) + H_2O \rightleftharpoons HA(aq) + OH^-(aq) \tag{7}$$

Solutions of sodium acetate, $NaC_2H_3O_2$, the salt formed by reaction of sodium hydroxide with acetic acid, will be slightly *basic* because of the reaction of $C_2H_3O_2^-$ ion with water to produce $HC_2H_3O_2$ and OH^-. Because of the analogous reaction of the NH_4^+ ion with water to form H_3O^+ ion, solutions of ammonium chloride, NH_4Cl, will be slightly *acidic*.

Salts of most transition metal ions are acidic. A solution of $CuSO_4$ or $FeCl_3$ will typically have a pH equal to 5 or lower. The salts are completely ionized in solution. The acidity comes from the fact that the cation is hydrated (e.g., $Cu(H_2O)_4{}^{2+}$ or $Fe(H_2O)_6{}^{3+}$). The large + charge on the metal cation attracts electrons from the O—H bonds in water, weakening them and producing some H^+ ions in solution; with $CuSO_4$ solutions the reaction would be

$$Cu(H_2O)_4{}^{2+}(aq) \rightleftharpoons Cu(H_2O)_3OH^+(aq) + H^+(aq) \tag{8}$$

Some solutions, called buffers, are remarkably resistant to pH changes caused by the addition of an acid or base. These solutions almost always contain both the salt of a weak acid or base and the parent acid or base. The solution used in the previously mentioned procedure for finding K_a of a weak acid is an example of a buffer solution. That solution contained equal amounts of the weak acid HA and the anion A^- present in its salt. If a small amount of strong acid were added to that solution, the H^+ ion would tend to react with the A^- ion present, keeping [H^+] about where it was before the addition. Similarly, a small amount of strong base added to that solution would react with the HA present, producing some A^- ion and water but not appreciably changing [OH^-]. If similar amounts of acid or base were added to *water* the pH could easily change by several units.

In this experiment you will first measure the pH of an unknown solution by using acid-base indicators. You will then use a pH meter to find the pH of some typical solutions. Finally, you will measure the dissociation constant of an unknown weak acid by the method described in the discussion. In the process you will prepare a buffer, whose properties will be investigated.

EXPERIMENTAL PROCEDURE

You may work in pairs on the first two parts of this experiment.

A. Determination of the pH of a Solution by Using Acid-Base Indicators. Obtain a solution of unknown pH from your laboratory assistant. Put about 1 cm³ of the solution into a small test tube and then add a drop or two of one of the indicators in Table 25.1. Note the color of the solution you obtain. Repeat the test with each of the indicators in the table. From the information in the table, estimate the pH of the solution roughly. Note that the color of an indicator is most indicative of the pH in the region where the indicator is changing color.

Table 25.1

Indicator	Approximate pH Range	Color Change
Methyl violet	0 to 2	yellow to violet
Thymol blue	1 to 3	red to yellow
Methyl yellow	2 to 4	red to yellow
Congo red	3 to 5	violet to red
Bromcresol green	3 to 6	yellow to blue

Having established the pH of your solution to within about one pH unit, obtain a known solution with a pH about equal to that of your unknown. Stock solutions with integral pH values from 0 to 6 have been prepared for your use. Test 1-ml portions of the known solution with the indicator or indicators that were most useful in fixing the pH of your unknown. Compare the colors of your known and unknown solutions. On the basis of the comparison, decide whether the pH of your unknown is higher or lower than that of the known. Select a second known solution, with a view toward bracketing your unknown between two known solutions. Treat that solution with the useful indicator(s) and compare its color(s) with that of your known. Continue testing known solutions until you find two which differ by one pH unit, one of which has a pH higher than that of the unknown and one with a pH that is lower. For the final estimation of pH, use 5-ml portions of your unknown and the two known solutions and two drops of indicator in each solution. By using a well-lit piece of white paper as a background, and viewing down the tubes, you should be able to estimate the pH of your unknown to ±0.3 pH units.

B. Measurement of the pH of Some Typical Solutions. In the rest of the experiment we will be using pH meters to find pH. Your instructor will show you how to operate your meter. The electrodes are fragile, so use due caution when using the pH meter.

Using a 25-ml sample in a 150-ml beaker, measure and record the pH of a 0.1 M solution of each of the following substances:

$$NaCl \quad HC_2H_3O_2 \quad Na_2CO_3$$
$$HCl \quad NH_3 \quad ZnCl_2$$

For each solution write a net ionic equation which explains, qualitatively, or, if possible, quantitatively, why the observed values of pH are reasonable.

C. Determination of the Dissociation Constant of a Weak Acid. In the rest of the experiment each student is to work independently.

Obtain from the stockroom a sample of an unknown acid and a buret. Measure out 100 ml of distilled water into a graduated cylinder and pour it into a clean 250-ml Erlenmeyer flask. Dissolve your acid sample in the water and stir thoroughly.

Pour half the solution into another 250-ml Erlenmeyer flask. Use the solution levels in the two flasks to decide when the volumes of solution in the two flasks are equal. Titrate the acid in one of the flasks to a phenolphthalein end point, using 0.2 M NaOH in the buret. (See Exp. 22; volume readings do not have to be taken here.) This should take less than 50 ml of the NaOH. Add the sodium hydroxide solution slowly while rotating the flask. As the end point approaches, add the solution drop by drop until the solution has a permanent pink color.

Mix the neutralized solution with the acid solution in the other flask and determine the pH of the resulting half-neutralized solution. From the observed pH calculate K_a for the unknown acid.

D. Properties of a Buffer. The solution you prepared in Part C, whose pH you measured, is a buffer. It contains a weak acid, HA, and its conjugate base, A^-, in equal concentrations. This solution is much more resistant to pH change than water would be, since the acid, HA, it contains can react with added OH^- ion, and the anion, A^-, can react with added H^+ ion.

To 25 ml of this buffer solution add 5 drops 0.1 M HCl and mix thoroughly. Measure the pH of the resulting solution. Measure the pH of the distilled water in the laboratory. To a 25-ml sample of distilled water add 5 drops of 0.1 M HCl. After mixing, measure the pH of the solution.

To another 25-ml sample of the buffer, add 5 drops 0.1 M NaOH. Mix well, and measure the pH of the resulting solution. Then add 5 drops 0.1 M NaOH to 25 ml of distilled water and measure the pH.

Write a net ionic equation showing why the buffer solution resists change in pH when H^+ ion is added. Write a second equation showing how the buffer reacts to resist pH change when OH^- ion is added. Finally, write an equation for the reaction producing the pH you observe when 0.1 M HCl is added to water.

OBSERVATIONS, CALCULATIONS, and EXPLANATIONS: pH, Its Measurement and Applications

A. Determination of pH Using Acid-Base Indicators

Indicator	Color with Unknown	Color with Known Solution
Methyl violet	_____	_____
Thymol blue	_____	_____
Methyl yellow	_____	_____
Congo red	_____	_____
Bromcresol green	_____	_____

Approx. pH of unknown _____

pH of first known solution _____

Indicator(s) used in final pH determination _____ _____

pH of second known solution _____

pH of third known solution (if needed) _____

Approximate pH of unknown solution ($\pm$ 0.3) _____

Unknown no. _____

B. Measurement of the pH of Some Typical Solutions

Record the pH of the 0.1 M solutions that were tested.

NaCl _____ $HC_2H_3O_2$ _____ Na_2CO_3 _____

HCl _____ NH_3 _____ $ZnCl_2$ _____

Continued on following page

Write a net ionic equation for the reaction responsible for the observed pH value for each solution.

NaCl _____ $HC_2H_3O_2$ _____

HCl _____ Na_2CO_3 _____

NH_3 _____ $ZnCl_2$ _____

C. Determination of the Dissociation Constant of a Weak Acid

pH of the half-neutralized acid solution   _____

$[H^+]$ in the half-neutralized solution   _____ M

K_a of the unknown acid   _____

Unknown no.   _____

D. Properties of an HA-A⁻ Buffer

	Buffer	Buffer + 5 drops 0.1 M HCl	Buffer + 5 drops 0.1 M NaOH
pH	_____	_____	_____

	Water	Water + 5 drops 0.1 M HCl	Water + 5 drops 0.1 M NaOH
pH	_____	_____	_____

Equation for reaction of buffer with H^+ ion: _____

Equation for reaction of buffer with OH^- ion: _____

Equation for reaction of HCl in water: _____

ADVANCE STUDY ASSIGNMENT: pH, Its Measurement and Applications

1.   A 0.5 M solution of an acid was tested with the indicators used in this experiment. The colors observed were as follows:

> Methyl violet: violet   Congo red: violet
> Thymol blue: yellow   Bromcresol green: yellow
> Methyl yellow: orange

What is the approximate pH of the solution?   _____

Which indicators would you use to find the pH more precisely?

_____   _____

2.   A solution of NaCN has a pH of 10. The CN^- ion is the anion of the weak acid HCN. Write the net ionic equation for the reaction which produces the pH of the NaCN solution (See Eq. 7).

3.   The pH of a half-neutralized solution of a weak acid is 4.8. What is the dissociation constant of the weak acid? (See Eq. 6)

$$K_a = \underline{\qquad\qquad}$$

4.   A solution with a volume of one liter contains 0.5 mole $HC_2H_3O_2$, 0.4 mole $C_2H_3O_2^-$ ion, and 0.4 mole Na^+ ion. K_a for $HC_2H_3O_2$ is 1.8×10^{-5}.

   a. What is the pH of the solution? (See Eq. 4)   _____

   b. Is the solution a buffer?   _____

   c. Why, or why not?

   d. Write the net ionic equation for the reaction which occurs if a little H^+ ion is added to the solution.

EXPERIMENT

26 • Relative Stabilities of Complex Ions and Precipitates Prepared from Solutions of Copper(II)

In aqueous solution, typical cations, particularly those produced from atoms of the transition metals, do not exist as free ions but rather consist of the metal ion in combination with some water molecules. Such cations are called complex ions. The water molecules, usually 2, 4, or 6 in number, are bound chemically to the metallic cation, but often rather loosely, with the electrons in the chemical bonds being furnished by one of the unshared electron pairs from the oxygen atoms in the H_2O molecules. Copper ion in aqueous solution may exist as $Cu(H_2O)_4^{2+}$, with the water molecules arranged in a square around the metal ion at the center.

If a hydrated cation such as $Cu(H_2O)_4^{2+}$ is mixed with other species that can, like water, form coordinate covalent bonds with Cu^{2+}, those species, called ligands, may displace one or more H_2O molecules and form other complex ions containing the new ligands. For instance, NH_3, a reasonably good coordinating species, may replace H_2O from the hydrated copper ion, $Cu(H_2O)_4^{2+}$, to form $Cu(H_2O)_3NH_3^{2+}$, $Cu(H_2O)_2(NH_3)_2^{2+}$, $Cu(H_2O)(NH_3)_3^{2+}$, or $Cu(NH_3)_4^{2+}$. At moderate concentrations of NH_3, essentially all the H_2O molecules around the copper ion are replaced by NH_3 molecules, forming the copper ammonia complex ion.

Coordinating ligands differ in their tendency to form bonds with metallic cations, so that in a solution containing a given cation and several possible ligands, an equilibrium will develop in which most of the cations are coordinated with those ligands with which they form the most stable bonds. There are many kinds of ligands, but they all share the common property that they possess an unshared pair of electrons which they can donate to form a coordinate covalent bond with a metal ion. In addition to H_2O and NH_3, other uncharged coordinating species include CO and ethylenediamine; some common anions that can form complexes include OH^-, Cl^-, CN^-, SCN^-, and $S_2O_3^{2-}$.

As you know, when solutions containing metallic cations are mixed with other solutions containing ions, precipitates are sometimes formed. When a solution of 0.1 M copper nitrate is mixed with a little 1 M NH_3 solution, a precipitate forms and then dissolves in excess ammonia. The formation of the precipitate helps us to understand what is occurring as NH_3 is added. The precipitate is hydrous copper hydroxide, formed by reaction of the hydrated copper ion with the small amount of hydroxide ion present in the NH_3 solution. The fact that this reaction occurs means that even at very low OH^- ion concentration $Cu(OH)_2(H_2O)_2(s)$ is a more stable species than $Cu(H_2O)_4^{2+}$ ion.

Addition of more NH_3 causes the solid to redissolve. The copper species then in solution cannot be the hydrated copper ion. (Why?) It must be some other complex ion, and is, indeed, the $Cu(NH_3)_4^{2+}$ ion. The implication of this reaction is that the $Cu(NH_3)_4^{2+}$ ion is also more stable in NH_3 solution than is the hydrated copper ion. To deduce in addition that the copper ammonia complex ion is also more stable in general than $Cu(OH)_2(H_2O)_2(s)$ is not warranted, since under the conditions in the solution $[NH_3]$ is much larger than $[OH^-]$, and given a higher concentration of hydroxide ion, the solid hydrous copper hydroxide might possibly precipitate even in the presence of substantial concentrations of NH_3.

To resolve this question, you might proceed to add a little 1 M NaOH solution to the solution containing the $Cu(NH_3)_4^{2+}$ ion. If you do this you will find that $Cu(OH)_2(H_2O)_2(s)$ does indeed precipitate. We can conclude from these observations that $Cu(OH)_2(H_2O)_2(s)$ is more stable than $Cu(NH_3)_4^{2+}$ in solutions in which the ligand concentrations (OH^- and NH_3) are roughly equal.

The copper species that will be present in a system depends, as we have just seen, on the conditions in the system. We cannot say in general that one species will be more stable than another; the stability of a given species depends in large measure on the kinds and concentrations of other species that are also present with it.

Another way of looking at the matter of stability is through equilibrium theory. Each of the copper species we have mentioned can be formed in a reaction between the hydrated copper ion and a complexing or precipitating ligand; each reaction will have an associated equilibrium constant, which we might call a formation constant for that species. The pertinent formation reactions and their constants for the copper species we have been considering are listed here:

$$Cu(H_2O)_4{}^{2+}(aq) + 4 NH_3(aq) \rightleftharpoons Cu(NH_3)_4{}^{2+}(aq) + 4 H_2O \qquad K_1 = 5 \times 10^{12} \qquad (1)$$

$$Cu(H_2O)_4{}^{2+}(aq) + 2 OH^-(aq) \rightleftharpoons Cu(OH)_2(H_2O)_2(s) + 2 H_2O \qquad K_2 = 2 \times 10^{19} \qquad (2)$$

The formation constants for these reactions do not involve $[H_2O]$ terms, which are essentially constant in aqueous systems and are included in the magnitude of K in each case. The large size of each formation constant indicates that the tendency for the hydrated copper ion to react with the ligands listed is very high.

In terms of these data, let us compare the stability of the $Cu(NH_3)_4{}^{2+}$ complex ion with that of solid $Cu(OH)_2(H_2O)_2$. This is most readily done by considering the reaction:

$$Cu(NH_3)_4{}^{2+}(aq) + 2 OH^-(aq) + 2 H_2O \rightleftharpoons Cu(OH)_2(H_2O)_2(s) + 4 NH_3(aq) \qquad (3)$$

We can find the value of the equilibrium constant for this reaction by noting that it is the sum of Reaction 2 and the reverse of Reaction 1. By the Law of Multiple Equilibrium, K for Reaction 3 is given by the equation

$$K = \frac{K_2}{K_1} = \frac{2 \times 10^{19}}{5 \times 10^{12}} = 4 \times 10^6 = \frac{[NH_3]^4}{[Cu(NH_3)_4{}^{2+}][OH^-]^2} \qquad (4)$$

From the expression in Equation 4 we can calculate that in a solution in which the NH_3 and OH^- ion concentrations are both about 1 M,

$$[Cu(NH_3)_4{}^{2+}] = \frac{1}{4 \times 10^6} = 2.5 \times 10^{-7} \text{ M} \qquad (5)$$

Since the concentration of the copper ammonia complex ion is very, very low, any copper(II) in the system will exist as the solid hydroxide. In other words, the solid hydroxide is more stable under such conditions than the ammonia complex ion. But that is exactly what we observed when we treated the hydrated copper ion with ammonia and then with an equivalent amount of hydroxide ion.

Starting now from the experimental behavior of the copper ion, we can conclude that since the solid hydroxide is the species that exists when copper ion is exposed to equal concentrations of ammonia and hydroxide ion, the hydroxide is more stable under those conditions, *and* the equilibrium constant for the formation of the hydroxide is larger than the constant for the formation of the ammonia complex. By determining, then, which species is present when a cation is in the presence of equal ligand concentrations, we can speak meaningfully of stability under such conditions and can rank the formation constants for the possible complex ions, and indeed for precipitates, in order of their increasing magnitudes.

In this experiment you will carry out formation reactions for a group of complex ions and precipitates involving the Cu^{2+} ion. You can make these species by mixing a solution of $Cu(NO_3)_2$ with solutions containing NH_3 or anions, which may form either precipitates or complex ions by reaction with $Cu(H_2O)_4{}^{2+}$, the cation present in aqueous solutions of copper(II) nitrate. By examining whether the precipitates or complex ions formed by the reaction of hydrated copper(II) ion with a given species can, on addition of a second ligand, be dissolved or transformed to another species, you will be able to rank the relative stabilities of the precipitates and complex ions made from Cu^{2+} with respect to one another, and thus rank the equilibrium formation constants for each species in order of increasing magnitude. The species to be reacted with Cu^{2+} ion in aqueous solution are NH_3, Cl^-, OH^-, CO_3^{2-}, $C_2O_4^{2-}$, S^{2-}, NO_2^-, and PO_4^{3-}. In each case the test for relative stability will be made in the presence of essentially equal concentrations of the two ligands. When you have completed your ranking of the known species you will test an unknown species and incorporate it into your list.

EXPERIMENTAL PROCEDURE

Obtain from the stockroom an unknown and enough small test tubes so that you have a total of eight.

Add about 2 ml 0.1 M $Cu(NO_3)_2$ solution to each of the test tubes.

To one of the test tubes add about 2 ml 1 M NH_3, drop by drop. Note whether a precipitate forms initially, and if it dissolves in excess NH_3. Shake the tube sideways to mix the reagents. In the NH_3-NH_3 space in the table (data page), write a P in the upper left hand corner if a precipitate is formed initially. In the rest of the space, write the formula and the color of the species which was present with an excess of NH_3. If a solution is present, the copper ion will be in a complex. Cu(II) will always have a coordination number of 4, so the formula with NH_3 would be $Cu(NH_3)_4^{2+}$. If a precipitate is present, it will be neutral, and in the case of NH_3 it would be a hydroxide with the formula $Cu(OH)_2$ (there should in principle be two H_2O molecules in the formula, but they are usually omitted). Add two ml 1 M NH_3 to the rest of the test tubes.

Now you will test the stability of the species present in excess NH_3 relative to those which might be present with other precipitating or coordinating species. Add, drop by drop, 2 ml of a 1 M solution of each of the anions in the horizontal row of the table to the test tubes you have just prepared, one solution to a test tube. Note any changes that occur in the appropriate spaces in the table. A change in color or the formation of a precipitate implies that a reaction has occurred between the added ligand or precipitating anion and the species originally present. As before, put a P in the upper left hand corner if a precipitate initially forms on addition of the anion solution. In the rest of the space, write the formula and color of the species present when the anion is in excess. That is the species which is stable in the presence of equal concentrations of NH_3 and the added anion. Again, in complex ions, Cu(II) will normally be attached to four ligands; copper(II) precipitates will be neutral. If a new species forms on addition of the second reagent, its formula should be given. If no change occurs, the original species is more stable, so put its formula in that space.

Repeat the above series of experiments, using 1 M Cl^- as the species originally added to the $Cu(NO_3)_2$ solution. In each case record the color of any precipitates or solutions formed on addition of the reagents in the horizontal row, and the formula of the species stable when an excess of both Cl^- ion and the added species is present in the solution. Since these reactions are reversible, it is not necessary to retest Cl^- with NH_3, since the same results would be obtained as when the NH_3 solution was tested with Cl^- solution.

Repeat the series of experiments for each of the anions in the vertical row in the table, omitting those tests where decisions as to relative stabilities are already clear. Where both ligands produce precipitates it will be helpful to check the effect of the addition of the other ligand to those precipitates. When complete, your table should have at least 36 entries.

Examine your table and decide on the relative stabilities of all species you observed to be present in all the spaces of the table. There should be eight such species; rank them as best you can in order of increasing stability. There is only one correct ranking, and you should be prepared to defend your choices. Although we did not in general prepare the species by direct reaction of $Cu(H_2O)_4^{2+}$ with the added ligand or precipitating anion, the equilibrium formation constants for those species for the direct reactions will have magnitudes that increase in the same order as the relative stabilities of the species you have established.

When you are satisfied that your ranking order is correct, carry out the necessary tests on your unknown to determine its proper position in the list. Your unknown may be one of the Cu(II) species you have already observed, or it may be a different species, present in excess of its ligand or precipitating anion. If your unknown contains a precipitate, shake it well before using a portion of it to make a test.

Name _____ Section _____

DATA AND OBSERVATIONS: Relative Stabilities of Complex Ions and Precipitates Containing Cu(II)

Table of Observations

	NH_3	Cl^-	OH^-	CO_3^{2-}	$C_2O_4^{2-}$	S^{2-}	NO_2^-	PO_4^{3-}	Unknown
PO_4^{3-}									
NO_2^-									
S^{2-}									
$C_2O_4^{2-}$									
CO_3^{2-}									
OH^-									
Cl^-									
NH_3									

Continued on following page   217

Determination of Relative Stabilities

In each row of the table you can compare the stabilities of species involving the reagent in the horizontal row with those of the species containing the reagent initially added. In the first row of the table, the copper(II)–NH_3 species can be seen to be more stable than some of the species obtained by addition of the other reagents, and less stable than others. Examining each row, make a list of all the complex ions and precipitates you have in the table in order of increasing stability and formation constant.

Reasons

Lowest _____   _____

_____   _____

_____   _____

_____   _____

_____   _____

_____   _____

Highest _____   _____

Stability of Unknown

Indicate the position your unknown would occupy in the above list.

Reasons:

Unknown no._____

ADVANCE STUDY ASSIGNMENT: Stabilities of Complex Ions and Precipitates of Cu(II)

1.   In testing the relative stabilities of Cu(II) species a student adds 2 ml 1 M NH_3 to 2 ml 0.1 M $Cu(NO_3)_2$. He observes that a blue precipitate initially forms, but that in excess NH_3 the precipitate dissolves and the solution turns blue. Addition of 2 ml 1 M NaOH to the dark blue solution results in the formation of a blue precipitate.

   a. What is the formula of Cu(II) species in the dark blue solution?

   b. What is the formula of the blue precipitate present after addition of 1 M NaOH?

   c. Which species is more stable in equal concentrations of NH_3 and OH^-, the one in Part a or the one in Part b?

   ———————

2.   Given the following two reactions and their equilibrium constants:

$$Cu(H_2O)_4^{2+}(aq) + 4\,NH_3(aq) \rightleftharpoons Cu(NH_3)_4^{2+}(aq) + 4\,H_2O \qquad K_1 = 5 \times 10^{12}$$

$$Cu(H_2O)_4^{2+}(aq) + CO_3^{2-} \rightleftharpoons CuCO_3(s) + 4\,H_2O \qquad K_2 = 7 \times 10^9$$

   a. Evaluate the equilibrium constant for the reaction

$$Cu(NH_3)_4^{2+}(aq) + CO_3^{2-} \rightleftharpoons CuCO_3(s) + 4\,NH_3$$

   ———————

   b. Given the value of K in part a and the relationship between stability and the formation constants, K_1 and K_2, find the concentration of $Cu(NH_3)_4^{2+}$ that would have to be present in a solution in which $[NH_3] = [CO_3^{2-}] = 1$ M and $CuCO_3$ was present as a solid. Interpret your results in words.

   ——————————M

27 • Determination of the Hardness of Water

One of the factors that establishes the quality of a water supply is its degree of hardness. The hardness of water is defined in terms of its content of calcium and magnesium ions. Since the analysis does not distinguish between Ca^{2+} and Mg^{2+}, and since most hardness is caused by carbonate deposits in the earth, hardness is usually reported as total parts per million calcium carbonate by weight. A water supply with a hardness of 100 parts per million would contain the equivalent of 100 grams of $CaCO_3$ in 1 million grams of water or 0.1 gram in one liter of water. In the days when soap was more commonly used for washing clothes, and when people bathed in tubs instead of using showers, water hardness was more often directly observed than it is now, since Ca^{2+} and Mg^{2+} form insoluble salts with soaps and make a scum that sticks to clothes or to the bath tub. Detergents have the distinct advantage of being effective in hard water, and this is really what allowed them to displace soaps for laundry purposes.

Water hardness can be readily determined by titration with the chelating agent EDTA (ethylenediaminetetraacetic acid). This reagent is a weak acid that can lose four protons on complete neutralization; its structural formula is

$$\begin{array}{c} HOOC-CH_2 \\ HOOC-CH_2 \end{array} N-CH_2-CH_2-N \begin{array}{c} CH_2-COOH \\ CH_2-COOH \end{array}$$

The four acid sites and the two nitrogen atoms all contain unshared electron pairs, so that a single EDTA ion can form a complex with up to six sites on a given cation. The complex is typically quite stable, and the conditions of its formation can ordinarily be controlled so that it contains EDTA and the metal ion in a $1:1$ mole ratio. In a titration to establish the concentration of a metal ion, the EDTA which is added combines quantitatively with the cation to form the complex. The end point occurs when essentially all of the cation has reacted.

In this experiment we will standardize a solution of EDTA by titration against a standard solution made from calcium carbonate, $CaCO_3$. We will then use the EDTA solution to determine the hardness of an unknown water sample. Since both EDTA and Ca^{2+} are colorless, it is necessary to use a rather special indicator to detect the end point of the titration. The indicator we will employ is called Calmagite, which forms a rather stable wine-red complex, $MgIn^-$, with the magnesium ion. A tiny amount of this complex will be present in the solution during the titration. As EDTA is added, it will complex free Ca^{2+} and Mg^{2+} ions, leaving the $MgIn^-$ complex alone until essentially all of the calcium and magnesium have been converted to chelates. At this point EDTA concentration will increase sufficiently to displace Mg^{2+} from the indicator complex; the indicator reverts to an acid form, which is sky blue, and this establishes the end point of the titration.

The titration is carried out at a pH of 10, in an NH_3–NH_4^+ buffer, which keeps the EDTA (H_4Y) mainly in the HY^{3-} form, where it complexes the Group 2 ions very well but does not tend to react as readily with other cations such as Fe^{3+} that might be present as impurities in the water. Taking H_4Y and H_3In as the formulas for EDTA and Calmagite respectively, the equations for the reactions which occur during the titration are:

$$\text{(main reaction) } HY^{3-}(aq) + Ca^{2+}(aq) \rightarrow CaY^{2-}(aq) + H^+(aq) \text{ (same for } Mg^{2+})$$

$$\text{(at end point) } \underset{\text{wine red}}{HY^{3-}(aq) + MgIn^-(aq)} \rightarrow MgY^{2-}(aq) + \underset{\text{sky blue}}{HIn^{2-}(aq)}$$

Since the indicator requires a trace of Mg^{2+} to operate properly, a little magnesium ion has been added to the stock EDTA solution.

EXPERIMENTAL PROCEDURE

WEAR YOUR SAFETY GLASSES WHILE
PERFORMING THIS EXPERIMENT

Obtain a 50-ml buret, a 250-ml volumetric flask, and 25- and 50-ml pipets from the stockroom.

Put about a half gram of calcium carbonate in a small 50 ml beaker and weigh the beaker and contents on the analytical balance. Using a spatula, transfer about 0.4 g of the carbonate to a 250-ml beaker and weigh again, determining the mass of the $CaCO_3$ sample by difference.

Add 25 ml of distilled water to the large beaker and then, *slowly*, about 40 drops of 6 M HCl. Cover the beaker with a watch glass and allow the reaction to proceed until all of the solid carbonate has dissolved. Rinse the walls of the beaker down with distilled water from your wash bottle and heat the solution until it just begins to boil. (Be sure not to be confused by the evolution of CO_2 which occurs with the boiling.) Add 50 ml of distilled water to the beaker and carefully transfer the solution, using a stirring rod as a pathway, to the volumetric flask. Rinse the beaker several times with small portions of distilled water and transfer each portion to the flask. All of the Ca^{2+} originally in the beaker should then be in the volumetric flask. Fill the volumetric flask to the horizontal mark with distilled water, adding the last few ml a drop at a time with your wash bottle. Stopper the flask and mix the solution thoroughly by inverting the flask at least a dozen times and shaking at intervals over a period of five minutes.

Clean your buret thoroughly and draw about 200 ml of the stock EDTA solution from the carboy into a dry 250-ml Erlenmeyer flask. Rinse the buret with a few ml of the solution at least three times. Drain through the stopcock and then fill the buret with the EDTA solution.

Prepare a reference solution by adding 50 ml distilled water and 5 ml of pH 10 buffer to a 125-ml Erlenmeyer flask. Add 8 drops of Calmagite indicator. The solution should turn sky blue. Use this solution as a reference for the end-point determination in your first titration.

Pipet a 25.00-ml aliquot of the Ca^{2+} solution into a clean 125-ml Erlenmeyer flask. Add 5 ml of the pH 10 buffer and 8 drops of Calmagite indicator, which will cause the solution to turn red. Read the level in the buret, and then add EDTA to the solution. Near the end-point the color may appear to fade, and will become more violet. The end-point occurs where the last tinge of red disappears and the solution takes on a color very much like that in the reference. The reaction at the end-point is rather slow ($\sim$1 sec), so take your time. Record the buret level at the end-point. Keep your titrated solution as a reference for your next titration. (If you are in doubt about the end-point color, add a few drops of the Ca^{2+} solution to the flask, and titrate again with the EDTA, recording no volumes but simply practicing until you can hit the end-point within one drop. The end-point is actually a good one, but experience with it does help.)

Refill your buret and titrate two more 25.00-ml aliquots of the calcium containing solution, using 5 ml of pH 10 buffer and 8 drops of indicator as before. With these titrations you should do somewhat better, since you have a good idea about how much EDTA will be needed.

Your instructor will furnish you a sample of water for hardness analysis. Since the concentration of Ca^{2+} is probably lower than that in the standard calcium solution you prepared, pipet 50 ml of the water sample for each titration. As before, add 8 drops of indicator and 5 ml of pH 10 buffer before titrating. Carry out as many titrations as necessary to obtain two volumes of EDTA that agree within about 3 per cent. If the volume of EDTA required in the first titration is low due to the fact that the water is not very hard, increase the volume of the water sample so that in succeeding titrations, it takes at least 20 ml of EDTA to reach the end point. If the volume of solution becomes inconveniently large, use a 250-ml Erlenmeyer flask for the titration and add 16 drops of indicator to increase the color intensity.

Name _____ Section _____

DATA AND CALCULATIONS: Determination of the Hardness of Water

Mass of beaker
plus $CaCO_3$ _____ g

Volume Ca^{2+}
solution prepared _____ ml

Mass of beaker
less sample _____ g

Molarity of Ca^{2+} _____ M

Mass of $CaCO_3$
sample _____ g

Moles Ca^{2+} in each
aliquot titrated _____ moles

Number of moles $CaCO_3$ in sample
(Formula mass = 100.1) _____ moles

Standardization of EDTA Solution

Titration:	I	II	III
Initial buret reading	_____ ml	_____ ml	_____ ml
Final buret reading	_____ ml	_____ ml	_____ ml
Volume of EDTA	_____ ml	_____ ml	_____ ml

Average volume of EDTA required to titrate Ca^{2+} _____ ml

Molarity of EDTA $= \dfrac{\text{no. moles } Ca^{2+} \text{ in aliquot} \times 1000}{\text{average volume EDTA required (ml)}} =$ _____ M

Continued on following page **223**

Determination of Water Hardness

Titration: I   II   III

Volume of water used   _____ml _____ml _____ml

Initial buret reading   _____ml _____ml _____ml

Final buret reading   _____ml _____ml _____ml

Volume of EDTA   _____ml _____ml _____ml

Average volume EDTA
per liter of water   _____ml

No. moles EDTA per liter water _____ = No. moles $CaCO_3$ per liter water

No. grams $CaCO_3$
per liter water   _____g   Water hardness
                                           (1 ppm = 1 mg/liter) _____ppm $CaCO_3$

Unknown No. _____

Name _____ Section _____

ADVANCE STUDY ASSIGNMENT: Determination of the Hardness of Water

1.   A 0.5118-g sample of $CaCO_3$ is dissolved in 6 M HCl, and the resulting solution is diluted to 250.0 ml in a volumetric flask.
   a. How many moles of $CaCO_3$ are in the sample? (formula mass = 100.1)

_____ moles

   b. What is the molarity of the Ca^{2+} ion in the 250 ml of solution?

_____ M

   c. How many moles of Ca^{2+} ion are in a 25.00-ml aliquot of the solution?

_____ moles

2.   25.00-ml aliquots of the solution from Problem 1 are titrated with EDTA to the Calmagite end-point. An aliquot is found to require 28.55 ml of the EDTA before the solution becomes sky blue.
   a. How many moles of EDTA are there in the volume that is used?

_____ moles

   b. What is the molarity of the EDTA solution?

_____ M

3.   A 100-ml sample of hard water is titrated with the EDTA solution in Problem 2. The volume of EDTA required to reach the end-point is 22.44 ml.
   a. How many moles of EDTA are used in the titration?

_____ moles

   b. How many moles of Ca^{2+} ion are there in the 100-ml water sample?

_____ moles

   c. If the Ca^{2+} ion comes from $CaCO_3$, how many moles of $CaCO_3$ are there in 1 liter of the hard water? How many grams $CaCO_3$ per liter?

_____ moles _____ grams

   d. If 1 ppm $CaCO_3$ = 1 mg per liter, what is the water hardness in ppm $CaCO_3$?

_____ ppm $CaCO_3$

EXPERIMENT

28 • Determination of an Equivalent Mass by Electrolysis

In Experiment 24 we determined the equivalent mass of an acid by titration of a known mass of the acid with a standardized solution of NaOH. There we defined an equivalent mass of acid to be the amount of acid in grams that could furnish one mole of H^+ ion.

Experimentally we find that the equivalent mass of an element can be related in a fundamental way to the chemical effects observed in that phenomenon known as *electrolysis*. As you know, some liquids, because they contain ions, will conduct an electric current. If the two terminals on a storage battery, or any other source of D.C. voltage, are connected through metal electrodes to a conducting liquid, an electric current will pass through the liquid and chemical reactions will occur at the two metal electrodes; in this process electrolysis is said to occur, and the liquid is said to be electrolyzed.

At the electrode connected to the *negative* pole of the battery, a *reduction* reaction will invariably be observed. In this reaction electrons will usually be accepted by one of the species present in the liquid, which, in the experiment we shall be doing, will be an aqueous solution. The species reduced will ordinarily be a metallic cation or the H^+ ion or possibly water itself; the reaction which is actually observed will be the one that occurs with the least expenditure of electrical energy, and will depend on the composition of the solution. In the electrolysis cell we shall study, the reduction reaction of interest will occur in a slightly acidic medium; hydrogen gas will be produced by the reduction of hydrogen ion:

$$2\,H^+(aq) + 2\,e^- \rightarrow H_2(g) \tag{1}$$

In this reduction reaction, which will occur at the negative pole, or *cathode*, of the cell, for every H^+ ion reduced *one* electron will be required, and for every molecule of H_2 formed, *two* electrons will be needed.

Ordinarily in chemistry we deal not with individual ions or molecules but rather with moles of substances. In terms of moles, we can say that, by Equation 1,

The reduction of one mole of H^+ ion requires one mole of electrons
The production of one mole of $H_2(g)$ requires two moles of electrons

A mole of electrons is a fundamental amount of electricity in the same way that a mole of pure substance is a fundamental unit of matter, at least from a chemical point of view. A mole of electrons is called a *faraday*, after Michael Faraday, who discovered the basic laws of electrolysis. The amount of a species which will react with a *mole* of *electrons*, or *one faraday*, is equal to the *equivalent mass* of that species. Since one faraday will reduce one mole of H^+ ion, we say that the equivalent mass of hydrogen is 1.008 grams, equal to the mass of one mole of H^+ ion (or one-half mole of $H_2(g)$). To form one mole of $H_2(g)$ one would have to pass two faradays through the electrolysis cell.

In the electrolysis experiment we will perform we will measure the volume of hydrogen gas produced under known conditions of temperature and pressure. By using the Ideal Gas Law we will be able to calculate how many moles of H_2 were formed, and hence how many faradays of electricity passed through the cell.

At the positive pole of an electrolysis cell (the metal electrode that is connected to the + terminal of the battery), an *oxidation* reaction will occur, in which some species will give up electrons. This reaction, which takes place at the *anode* in the cell, may involve again an ionic or neutral species in the solution or the metallic electrode itself. In the cell that you will be studying, the pertinent oxidation reaction will be that in which a metal under study will participate:

$$M(s) \rightarrow M^{n+}(aq) + ne^- \tag{2}$$

During the course of the electrolysis the atoms in the metal electrode will be converted to metallic cations and will go into the solution. The mass of the metal electrode will decrease, depending on the amount of electricity passing through the cell and the nature of the metal. In order to oxidize one mole, or one molar mass, of the metal, it would take n faradays, where n is the charge on the cation which is formed. By definition, one faraday of electricity would cause one equivalent mass, EM, of metal to go into solution. The molar mass, MM, and the equivalent mass of the metal are related by the equation:

$$MM = EM \times n \tag{3}$$

In an electrolysis experiment, since n is not determined independently, it is not possible to find the molar mass of a metal. It is possible, however, to find equivalent masses of many metals, and that will be our main purpose.

The general method we will use is implied by the discussion. We will oxidize a sample of an unknown metal at the positive pole of an electrolysis cell, weighing the metal before and after the electrolysis and so determining its loss in mass. We will use the same amount of electricity, the same number of electrons, to reduce hydrogen ion at the negative pole of the electrolysis cell. From the volume of H_2 gas which is produced under known conditions we can calculate the number of moles of H_2 formed, and hence the number of faradays which passed through the cell. The equivalent mass of the metal is then calculated as the amount of metal which would be oxidized if one faraday were used. In an optional part of the experiment, your instructor may tell you the nature of the metal you used. Using Equation 3, it will be possible to determine the charge on the metallic cations that were produced during electrolysis.

EXPERIMENTAL PROCEDURE

Obtain from the stockroom a buret and a sample of a metal unknown. Lightly sand the metal to clean it. Weigh the metal sample on the analytical balance to 0.0001 g.

Set up the electrolysis apparatus as indicated in Figure 28.1. There should be about 100 ml 0.1 M $HC_2H_3O_2$ in 0.5 M Na_2SO_4 in the beaker with the gas buret. This will serve as the conducting solution. Immerse the end of the buret in the solution and attach a length of rubber tubing to its upper end. Open the stopcock on the buret and, with suction, carefully draw the acid up to the top of the graduations. Close the stopcock. Insert the bare coiled end of the heavy copper wire up into the end of the buret; all but the coil end of the wire should be covered with watertight insulation. Check the solution level after a few minutes to make sure the stopcock does not leak. Record the level.

The metal unknown will serve as the anode in the electrolysis cell. Connect the metal to the + pole of the power source with an alligator clip and immerse the metal but not the clip in the conducting solution. The copper electrode will be the cathode in the cell. Connect that electrode to the − pole of the power source. Hydrogen gas should immediately begin to bubble from the copper cathode. Collect the gas until about 50 ml have been produced. At that point, stop the electrolysis by disconnecting the copper electrode from the power source. Record the level of the liquid in the buret. Measure and record the temperature of the solution in the beaker and the barometric pressure in the laboratory. (In some cases a cloudiness may develop in the solution during the electrolysis. This is caused by the formation of a metal hydroxide, and will have no adverse effect on the experiment.)

Raise the buret, and discard the conducting solution in the beaker. Rinse the beaker with water, and pour in 100 ml of fresh conducting solution. Repeat the electrolysis, generating about

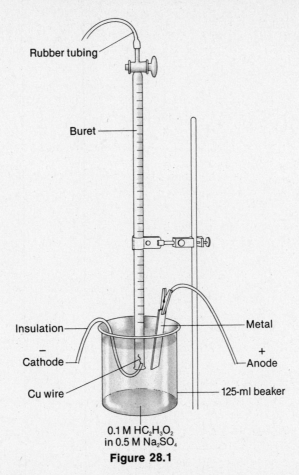

Rubber tubing

Buret

Insulation

– Cathode

Cu wire

Metal

+ Anode

125-ml beaker

0.1 M HC$_2$H$_3$O$_2$
in 0.5 M Na$_2$SO$_4$

Figure 28.1

50 ml of H$_2$ and recording the initial and final liquid levels in the buret. Take the alligator clip off the metal anode and wash the anode with 0.1 M HC$_2$H$_3$O$_2$, acetic acid. Rub off any loose adhering coating with your fingers, and then rinse off your hands. Rinse the electrode in water and then in acetone. Let the acetone evaporate. Weigh the dry metal electrode to the nearest 0.0001 g.

DATA AND CALCULATIONS: Determination of an Equivalent Mass by Electrolysis

Mass of metal anode   _____ g

Mass of anode after electrolysis   _____ g

Initial buret reading   _____ ml

Buret reading after first electrolysis   _____ ml

Buret reading after refilling   _____ ml

Buret reading after second electrolysis   _____ ml

Barometric pressure   _____ mm Hg

Temperature t _____ °C

Vapor pressure of H_2O at t _____ mm Hg

Total volume of H_2 produced, V _____ ml

Temperature T _____ K

Pressure exerted by dry H_2: $P = P_{Bar} - VP_{H_2O}$
(ignore any pressure effect due to liquid levels in buret)   _____ mm Hg

No. moles H_2 produced, n
(use Ideal Gas Law, $PV = nRT$)   _____ moles

No. of faradays passed   _____

Loss in mass by anode   _____ g

Equivalent mass of metal $\left(EM = \dfrac{\text{no. g lost}}{\text{no. faradays passed}}\right)$ _____ g

Unknown metal number   _____

Optional: Nature of metal _____ MM _____ g

Charge n on cation _____ (Eq. 3)

ADVANCE STUDY ASSIGNMENT: Determination of an Equivalent
Mass by Electrolysis

1. In an electrolysis cell similar to the one employed in this experiment, a student observed that his unknown metal anode lost 0.208 g while a total volume of 96.30 ml of H_2 was being produced. The temperature in the laboratory was 25°C and the barometric pressure was 748 mm Hg. At 25°C the vapor pressure of water is 23.8 mm Hg. To find the equivalent mass of his metal, he filled in the blanks below. Fill in the blanks as he did.

$P_{H_2} = P_{Bar} - VP_{H_2O} =$ _____ mm Hg = _____ atm

$V_{H_2} =$ _____ ml = _____ liters

$T =$ _____ K

$n_{H_2} =$ _____ moles $n_{H_2} = \dfrac{PV}{RT}$

1 mole H_2 requires passage of _____ faradays

No. of faradays passed = _____

Loss of mass of metal anode = _____ grams

No. grams of metal lost per faraday passed = $\dfrac{\text{no. grams lost}}{\text{no. faradays passed}} =$ _____ g = EM

The student was told that his metal anode was made of iron.

MM Fe = _____ g The charge n on the Fe ion is therefore _____ (Eq. 3)

EXPERIMENT

29 • Voltaic Cell Measurements

Many chemical reactions can be classified as oxidation-reduction reactions, since they involve the oxidation of one species and the reduction of another. Such reactions can conveniently be considered as the result of two half-reactions, one of oxidation and the other reduction. In the case of the oxidation-reduction reaction

$$Zn(s) + Pb^{2+}(aq) \rightarrow Zn^{2+}(aq) + Pb(s)$$

which would occur if a piece of metallic zinc were put into a solution of lead nitrate, the two reactions would be

$$Zn(s) \rightarrow Zn^{2+}(aq) + 2 \, e^- \qquad \text{oxidation}$$

$$2 \, e^- + Pb^{2+}(aq) \rightarrow Pb(s) \qquad \text{reduction}$$

The tendency for an oxidation-reduction reaction to occur can be measured if the two reactions are made to occur in separate regions connected by a barrier that is porous to ion movement. An apparatus called a *voltaic cell* in which this reaction might be carried out under this condition is shown in Figure 29.1.

If we connect a voltmeter between the two electrodes we will find that there is a voltage, or potential, between them. The magnitude of the potential is a direct measure of the driving force or thermodynamic tendency of the spontaneous oxidation-reduction reaction to occur.

If we study several oxidation-reduction reactions we find that the voltage of each associated voltaic cell can be considered to be the sum of a potential for the oxidation reaction and a potential for the reduction reaction. In the Zn, $Zn^{2+} \| Pb^{2+}$, Pb cell we have been discussing, for example,

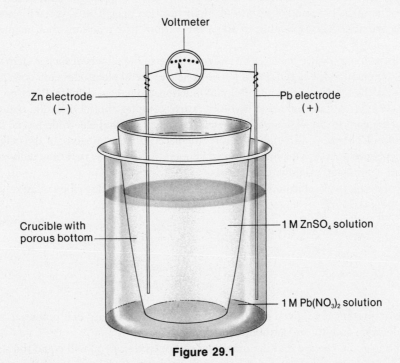

Figure 29.1

Voltmeter

Zn electrode
(−)

Pb electrode
(+)

Crucible with
porous bottom

1 M ZnSO₄ solution

1 M Pb(NO₃)₂ solution

$$E_{\text{cell}} = E_{\text{Zn, Zn}^{2+}\text{ oxidation reaction}} + E_{\text{Pb}^{2+}\text{, Pb reduction reaction}} \tag{1}$$

By convention, the negative electrode in a voltaic cell is taken to be the one from which electrons are emitted (i.e., where oxidation occurs). The negative electrode is the one which is connected to the minus pole of the voltmeter when the voltage is measured.

Since any cell potential is the sum of two electrode potentials, it is not possible, by measuring cell potentials, to determine individual absolute electrode potentials. However, if a value of potential is arbitrarily assigned to one electrode reaction, then other electrode potentials can be given definite values, based on the assigned value. The usual procedure is to assign a value of 0.0000 volts to the standard potential for the electrode reaction

$$2\,H^+(aq) + 2\,e^- \rightarrow H_2(g); \quad E^0_{H^+,\,H_2\,\text{red}} = 0.0000\ V$$

For the Zn, $Zn^{2+} \| H^+$, H_2 cell, the measured potential is 0.76 V, and the zinc electrode is negative. Zinc metal is therefore oxidized, and the cell reaction must be

$$Zn(s) + 2\,H^+(aq) \rightarrow Zn^{2+}(aq) + H_2(g); \quad E^0_{\text{cell}} = 0.76\ V$$

Given this information, one can readily find the potential for the oxidation of Zn to Zn^{2+}.

$$E^0_{\text{cell}} = E^0_{\text{Zn, Zn}^{2+}\text{ oxid}} + E^0_{H^+,\,H_2\,\text{red}}$$

$$0.76\ V = E^0_{\text{Zn, Zn}^{2+}\text{ oxid}} + 0.00\ V; \quad E^0_{\text{Zn, Zn}^{2+}\text{ oxid}} = +0.76\ V$$

If the potential for a half-reaction is known, the potential for the reverse reaction can be obtained by changing the sign. For example:

$$\textit{if } E^0_{\text{Zn, Zn}^{2+}\text{ oxid}} = +0.76 \text{ volts}, \quad \textit{then } E^0_{\text{Zn}^{2+}\text{, Zn red}} = -0.76 \text{ volts}$$

$$\textit{if } E^0_{\text{Pb}^{2+}\text{, Pb red}} = +Y \text{ volts}, \quad \textit{then } E^0_{\text{Pb, Pb}^{2+}\text{ oxid}} = -Y \text{ volts}$$

In the first part of this experiment you will measure the voltages of several different cells. By arbitrarily assigning the potential of a particular half-reaction to be 0.00 V, you will then be able to calculate the potentials corresponding to all of the various half reactions that occurred in your cells.

In our discussion so far we have not considered the possible effects of such system variables as temperature, potential at the liquid-liquid junction, size of metal electrodes, and concentrations of solute species. Although temperature and liquid junctions do have a definite effect on cell potentials, taking account of their influence involves rather complex thermodynamic concepts and is usually not of concern in any elementary course. The size of a metal electrode has no appreciable effect on electrode potential, although it does relate directly to the capacity of the cell to produce useful electrical energy. In this experiment we will operate the cells so that they deliver essentially no energy but exert their maximum potentials.

The effect of solute ion concentrations is important and can be described relatively easily. For the cell reaction at 25°C:

$$aA(s) + bB^+(aq) \rightarrow cC(s) + dD^{2+}(aq)$$

$$E_{\text{cell}} = E^0_{\text{cell}} - \frac{0.06}{n} \log \frac{(\text{conc. D}^{2+})^d}{(\text{conc. B}^+)^b} \tag{2}$$

where E^0_{cell} is a constant for a given reaction and is called the standard cell potential, and n is the number of electrons in either electrode reaction.

By Equation 2 you can see that the measured cell potential, E_{cell}, will equal the standard cell potential if the molarities of D^{2+} and B^+ are both unity, or, if d equals b, if the molarities are simply equal to each other. We will carry out our experiments under such conditions that the cell potentials you observe will be very close to the standard potentials given in the tables in your chemistry text.

Considering the $Cu,Cu^{2+}\|Ag^+,Ag$ cell as a specific example, the observed cell reaction would be

$$Cu(s) + 2\ Ag^+(aq) \rightarrow Cu^{2+}(aq) + 2\ Ag(s)$$

For this cell, Equation 2 takes the form

$$E_{cell} = E^0_{cell} - \frac{0.06}{2} \log \frac{conc.\ Cu^{2+}}{(conc.\ Ag^+)^2} \tag{3}$$

In the equation n is 2 because in the cell reaction, two electrons are transferred in each of the two half-reactions. E^0 would be the cell potential when the copper and silver salt solutions are both 1 M, since then the logarithm term is equal to zero.

If we decrease the Cu^{2+} concentration, keeping that of Ag^+ at 1 M, the potential of the cell will go up by about 0.03 volts for every factor of ten by which we decrease conc. Cu^{2+}. Ordinarily it is not convenient to change concentrations of an ion by several orders of magnitude, so in general, concentration effects in cells are relatively small. However, if we should add a complexing or precipitating species to the copper salt solution, the value of conc. Cu^{2+} would drop drastically, and the voltage change would be appreciable. In the experiment we will illustrate this effect by using NH_3 to complex the Cu^{2+}. Using Equation 3, we can actually calculate conc. Cu^{2+} in the solution of its complex ion.

In an analogous experiment we will determine the solubility product of AgCl. In this case we will surround the Ag electrode in a $Cu,Cu^{2+}\|Ag^+,Ag$ cell with a solution of known Cl^- ion concentration which is saturated with AgCl. From the measured cell potential, we can use Equation 3 to calculate the very small value of conc. Ag^+ in the chloride-containing solution. Knowing the concentrations of Ag^+ and Cl^- in a solution in equilibrium with AgCl(s) allows us to find K_{sp} for AgCl.

EXPERIMENTAL PROCEDURE

You may work in pairs in this experiment.

A. Cell Potentials. In this experiment you will be working with these seven electrode systems:

$Ag^+, Ag(s)$ $Br_2(l), Br^-, Pt$
$Cu^{2+}, Cu(s)$ $Cl_2(g, 1\ atm), Cl^-, Pt$
Fe^{3+}, Fe^{2+}, Pt $I_2(s), I^-, Pt$
$Zn^{2+}, Zn(s)$

Your purpose will be to measure enough voltaic cell potentials to allow you to determine the electrode potentials of each electrode by comparing it with an arbitrarily chosen electrode potential.

Using the apparatus shown in Figure 29.1, set up a voltaic cell involving any two of the electrodes in the list. About 10 ml of each solution should be enough for making the cell. The solute ion concentrations may be assumed to be one molar and all other species may be assumed to be at unit activity, so that the potentials of the cells you set up will be essentially the standard potentials. Measure the cell potential and record it along with which electrode has negative polarity.

In a similar manner set up and measure the cell voltage and polarities of other cells, sufficient in number to include all of the electrode systems on the list at least once. Do not combine the silver electrode system with any of the halogen electrode systems, since a precipitate will form; any other combinations may be used. The data from this part of the experiment should be entered in the first three columns of the table in Part A.1 of your report.

B.   Effect of Concentration on Cell Potentials

1.   COMPLEX ION FORMATION.   Set up the Cu,Cu^{2+}∥Ag$^+$,Ag cell, using 10 ml of the CuSO$_4$ solution in the crucible and 10 ml of AgNO$_3$ in the beaker. Measure the potential of the cell. While the potential is being measured, add 10 ml of 6 M NH$_3$ to the CuSO$_4$ solution, stirring carefully with your stirring rod. Measure the potential when it becomes steady.

2.   DETERMINATION OF THE SOLUBILITY PRODUCT OF AgCl. Remove the crucible from the cell you have just studied and discard the Cu(NH$_3$)$_4$$^{2+}$. Clean the crucible by drawing a little 6 M NH$_3$ through it, using the adapter and suction flask. Then draw some distilled water through it. Reassemble the Cu-Ag cell, this time using the beaker for the Cu-CuSO$_4$ electrode system. Immerse the Ag electrode in the crucible in 1 M KCl; add a drop of AgNO$_3$ solution to form a little AgCl, so that an equilibrium between Ag$^+$ and Cl$^-$ can be established. Measure the potential of this cell, noting which electrode is negative. In this case conc. Ag$^+$ will be very low, which will decrease the potential of the cell to such an extent that its polarity may change from that observed previously.

DATA AND CALCULATIONS:　Voltaic Cell Measurements

A.1.　Cell Potentials

Electrode systems used in cell	Cell potential, E^0_{cell} (volts)	Negative electrode	Oxidation reaction	$E^0_{oxidation}$ in volts	Reduction reaction	$E^0_{reduction}$ in volts
1.						
2.						
3.						
4.						
5.						
6.						
7.						

CALCULATIONS

A. Noting that oxidation occurs at the *negative* pole in a cell, write the oxidation reaction in each of the cells. The other electrode system must undergo reduction; write the reduction reaction which occurs in each cell.

B. Assume that $E^0_{Ag^+, Ag} = 0.00$ volts (whether in reduction or oxidation). Enter that value in the table for all of the silver electrode systems you used in your cells. Since $E^0_{cell} = E^0_{oxidation} + E^0_{reduction}$, you can calculate E^0 values for all the electrode systems in which the Ag, Ag$^+$ system was involved. Enter those values in the table.

C. Using the values and relations in B and taking advantage of the fact that for any given electrode system, $E^0_{oxidation} = -E^0_{reduction}$, complete the table of E^0 values. The best way to do this is to use one of the E^0 values you found in B in another cell with that electrode system. That potential, along with E^0_{cell}, will allow you to find the potential of the other electrode. Continue this process with other cells until all the electrode potentials have been determined.

Continued on following page

A.2.   Table of Electrode Potentials

In Table A.1, you should have a value for E^0_{red} or E^0_{oxid} for each of the electrode systems you have studied. Remembering that for any electrode system, $E^0_{red} = -E^0_{oxid}$, you can find the value for E^0_{red} for each system. List those potentials in the left column of the table below in order of decreasing value.

$E^0_{reduction}$ ($E^0_{Ag^+, Ag} = 0.00$ volts)	Electrode reaction in reduction	$E^0_{reduction}$ ($E^0_{H^+, H_2} = 0.00$ volts)
_____	_____	_____
_____	_____	_____
_____	_____	_____
_____	_____	_____
_____	_____	_____
_____	_____	_____
_____	_____	_____

The electrode potentials you have determined are based on $E^0_{Ag^+, Ag} = 0.00$ volts. The usual assumption is that $E^0_{H^+, H_2} = 0.00$ volts, under which conditions $E^0_{Ag^+, Ag \, red} = 0.80$ volts. Convert from one base to the other by adding 0.80 volts to each of the electrode potentials and enter these values in the third column of the table.

Why are the values of E^0_{red} on the two bases related to each other in such a simple way?

B.   Effect of Concentration on Cell Potentials

1.   Complex ion formation:

Potential, E^0_{cell}, before addition of 6 M NH_3 _____ volts

Potential, E_{cell}, after $Cu(NH_3)_4^{2+}$ formed _____ volts

Given Equation 3

$$E_{cell} = E^0_{cell} - \frac{0.06}{2} \log \frac{\text{conc. } Cu^{2+}}{(\text{conc. } Ag^+)^2} \tag{3}$$

Continued on following page

calculate the residual concentration of free Cu^{2+} ion in equilibrium with $Cu(NH_3)_4^{2+}$ in the solution in the crucible. Take conc. Ag^+ to be 1 M.

conc. Cu^{2+} = _____ M

2. Solubility product of AgCl:

Potential, E^0_{cell}, of the Cu, $Cu^{2+} \| Ag^+$, Ag cell (from B.1)

_____ V Negative electrode _____

Potential, E_{cell}, with 1 M KCl present   _____ V Negative electrode _____

Using Equation 3, calculate (conc. Ag^+) in the cell, where it is in equilibrium with 1 M Cl^- ion. (E_{cell} in Equation 3 is the *negative* of the measured value if the polarity is not the same as in the standard cell.) Take conc. Cu^{2+} to be 1 M.

conc. Ag^+ = _____ M

Since Ag^+ and Cl^- in the cell are in equilibrium with AgCl, we can find K_{sp} for AgCl from the concentration of Ag^+ and Cl^-, as they exist in the cell. Formulate the expression for K_{sp} for AgCl, and determine its value.

K_{sp} = _____

ADVANCE STUDY ASSIGNMENT: Voltaic Cell Measurements

1. A student measures the potential of a cell made up with 1 M $CuSO_4$ in one solution and 1 M $AgNO_3$ in the other. There is a Cu electrode in the $CuSO_4$ and an Ag electrode in the $AgNO_3$, and the cell is set up as in Figure 29.1. She finds that the potential, or voltage, of the cell, E^0_{cell}, is 0.47 volts, and that the Cu electrode is negative.

    a. At which electrode is oxidation occurring?   _____

    b. Write the equation for the oxidation reaction.

    c. Write the equation for the reduction reaction.

    d. If the potential of the silver, silver ion electrode, $E^0_{Ag^+,Ag}$ is taken to be 0.000 volts in oxidation or reduction, what is the value of the potential for the oxidation reaction, $E^0_{Cu,Cu^{2+} oxid}$? $E^0_{cell} = E^0_{oxid} + E^0_{red}$.

                                                          _____ volts

    e. If $E^0_{Ag^+,Ag red}$ equals 0.80 volts, as in standard tables of electrode potentials, what is the value of the potential of the oxidation reaction of copper, $E^0_{Cu,Cu^{2+} oxid}$?

                                                          _____ volts

    f. Write the net ionic equation for the spontaneous reaction that occurs in the cell which the student studied.

    g. The student adds 6 M NH_3 to the $CuSO_4$ solution until the Cu^{2+} ion is essentially all converted to $Cu(NH_3)_4^{2+}$ ion. The voltage of the cell, E_{cell}, goes up to 0.88 volts and the Cu electrode is still negative. Find the residual concentration of Cu^{2+} ion in the cell. Use Eq. 3.

                                                          _____ M

EXPERIMENT

30 • Preparation of Copper(I) Chloride

Oxidation-reduction reactions are, like precipitation reactions, often used in the preparation of inorganic substances. In this experiment we will employ a series of such reactions to prepare one of the less commonly encountered salts of copper, copper(I) chloride. Most copper compounds contain copper(II), but copper(I) is present in a few slightly soluble or complex copper salts.

The process of synthesis of CuCl we will use begins by dissolving copper metal in nitric acid:

$$Cu(s) + 4\,H^+(aq) + 2\,NO_3^-(aq) \rightarrow Cu^{2+}(aq) + 2\,NO_2(g) + 2\,H_2O \tag{1}$$

The solution obtained is treated with sodium carbonate in excess, which neutralizes the remaining acid with evolution of CO_2 and precipitates Cu(II) as the carbonate:

$$2\,H^+(aq) + CO_3^{2-}(aq) \rightleftharpoons (H_2CO_3)\,(aq) \rightleftharpoons CO_2(g) + H_2O \tag{2}$$

$$Cu^{2+}(aq) + CO_3^{2-}(aq) \rightleftharpoons CuCO_3(s) \tag{3}$$

The $CuCO_3$ will be purified by filtration and washing and dissolved in hydrochloric acid. Copper metal added to the highly acidic solution then reduces the Cu(II) to Cu(I) and is itself oxidized to Cu(I) in a disproportionation reaction. In the presence of excess chloride, the copper will be present as a $CuCl_4^{3-}$ complex ion. Addition of this solution to water destroys the complex, and white CuCl precipitates.

$$CuCO_3(s) + 2\,H^+(aq) + 4\,Cl^-(aq) \rightarrow CuCl_4^{2-}(aq) + CO_2(g) + H_2O \tag{4}$$

$$CuCl_4^{2-}(aq) + Cu(s) + 4\,Cl^-(aq) \rightarrow 2\,CuCl_4^{3-}(aq) \tag{5}$$

$$CuCl_4^{3-}(aq) \xrightarrow{H_2O} CuCl(s) + 3\,Cl^-(aq) \tag{6}$$

Since CuCl is readily oxidized, due care must be taken to minimize its exposure to air during its preparation and while it is being dried.

EXPERIMENTAL PROCEDURE

Obtain a 1 g sample of copper metal, a Buchner funnel, and a filter flask from the stockroom. Weigh the copper metal on the top loading or triple beam balance to 0.1 g.

Put the metal in a 150 ml beaker and *under a hood* add 5 ml 15 M HNO_3. **CAUTION: Caustic reagent.** Brown NO_2 gas will be evolved and an acidic blue solution of $Cu(NO_3)_2$ produced. If it is necessary, you may warm the beaker gently with a Bunsen burner to dissolve all of the copper. When all of the copper is in solution, add 50 ml of water to the solution and allow it to cool.

Weigh out about 5 grams of sodium carbonate in a small beaker on a rough balance. Add small amounts of the Na_2CO_3 to the solution with your spatula, adding the solid as necessary when the evolution of CO_2 subsides. Stir the solution to expose it to the solid. When the acid is neutralized, a blue-green precipitate of $CuCO_3$ will begin to form. At that point, add the rest of the Na_2CO_3, stirring the mixture well to ensure complete precipitation of the copper carbonate.

Transfer the precipitate to the Buchner funnel and use suction to remove the excess liquid.

Use your rubber policeman and a spray from your wash bottle to make a complete transfer of the solid. Wash the precipitate well with distilled water with suction on, then let it remain on the filter paper with suction on for a minute or two.

Remove the filter paper from the funnel and transfer the solid $CuCO_3$ to the 150 ml beaker. Add 10 ml water then 30 ml 6 M HCl slowly to the solid, stirring continuously. When the $CuCO_3$ has all dissolved, add 1.5 g Cu foil cut in small pieces to the beaker and cover it with a watch glass.

Heat the mixture in the beaker to the boiling point and keep it at that temperature, just simmering, for 30 to 40 minutes. It may be that the dark-colored solution which forms will clear to a yellow color before that time is up, and if it does, you may stop heating and proceed with the next step.

While the mixture is heating, put 150 ml distilled water in a 400 ml beaker and put the beaker in an ice bath. Cover the beaker with a watch glass. After you have heated the acidic Cu-$CuCl_2$ mixture for 40 minutes or as soon as it turns light-colored, carefully decant the hot liquid into the beaker of water, taking care not to transfer any of the excess Cu metal to the beaker. White crystals of CuCl should form. Continue to cool the beaker in the ice bath to promote crystallization and to increase the yield of solid.

Cool 25 ml distilled water, to which you have added 5 drops 6 M HCl, in an ice bath. Put 20 ml acetone into a small beaker. Filter the crystals of CuCl in the Buchner funnel using suction. Swirl the beaker to aid in transferring the solid to the funnel. Just as the last of the liquid is being pulled through, wash the CuCl with one third of the acidified cold water. Rinse the last of the CuCl into the funnel with another portion of the water and use the final third to rewash the solid. Turn off suction and add one half of the acetone to the funnel; wait about ten seconds and turn on the suction. Repeat this operation with the other half of the acetone. Draw air through the sample for a few minutes to dry it. If you have properly washed the solid, it will be pure white; if the moist compound is allowed to come into contact with air, it will tend to turn pale green, due to oxidation of Cu(I) to Cu(II). Weigh the CuCl in a previously weighed beaker to 0.1 g. Show your sample to your instructor for evaluation.

DATA AND RESULTS: Preparation of CuCl

Mass of Cu sample   ——————— g

Mass of beaker   ——————— g

Mass of beaker plus CuCl   ——————— g

Mass of CuCl prepared   ——————— g

Theoretical yield   ——————— g

Percentage yield   ——————— %

ADVANCED STUDY ASSIGNMENT: Preparation of Copper(I) Chloride

1. The Cu^{2+} ions in this experiment are produced from the reaction of 1.0 g of copper foil with excess nitric acid. How many moles of Cu^{2+} are produced?

_____ moles Cu^{2+}

2. Why isn't hydrochloric acid used in a direct reaction with copper wire to prepare the $CuCl_2$ solution?

3. How many grams of metallic copper are required to react with the number of moles of Cu^{2+} calculated in (1) to form the CuCl? The overall reaction can be taken to be: $Cu^{2+}(aq) + 2\,Cl^-(aq) + Cu(s) \rightarrow 2\,CuCl(s)$

_____ g Cu

4. What is the maximum mass of CuCl that can be prepared from the reaction sequence of this experiment using 1.0 g of Cu foil to prepare the Cu^{2+} solution?

_____ g CuCl

5. Indicate all the species, in proper sequence, to which the copper in the foil is transformed during the course of the synthesis of CuCl.

EXPERIMENT

31 • Determination of Iron by Reaction with Permanganate—A Redox Titration

Potassium permanganate, $KMnO_4$, is widely used as an oxidizing agent in volumetric analysis. In acid solution, MnO_4^- ion undergoes reduction to Mn^{2+} as shown in the following equation:

$$8\,H^+(aq) + MnO_4^-(aq) + 5\,e^- \rightarrow Mn^{2+}(aq) + 4\,H_2O$$

Since the MnO_4^- ion is violet and the Mn^{2+} ion is nearly colorless, the end point in titrations using $KMnO_4$ as the titrant can be taken as the first permanent pink color that appears in the solution.

$KMnO_4$ will be employed in this experiment to determine the percentage of iron in an unknown containing iron(II) ammonium sulfate, $Fe(NH_4)_2(SO_4)_2 \cdot 6\,H_2O$. The titration, which involves the oxidation of Fe^{2+} ion to Fe^{3+} by permanganate ion, is carried out in sulfuric acid solution to prevent the air-oxidation of Fe^{2+}. The end point of the titration is sharpened markedly if phosphoric acid is present, since the Fe^{3+} ion produced in the titration forms an essentially colorless complex with the acid.

The number of moles of potassium permanganate used in the titration is equal to the product of the molarity of the $KMnO_4$ and the volume used. The number of moles of iron present in the sample is obtained from the balanced equation for the reaction and the amount of MnO_4^- ion reacted. The per cent by weight of iron in the solid sample follows directly.

EXPERIMENTAL PROCEDURE

WEAR YOUR SAFETY GLASSES WHILE PERFORMING THIS EXPERIMENT

Obtain from the stockroom a buret and an unknown iron(II) sample.

Weigh out accurately on the analytical balance three samples of about 1.0 g of your unknown into clean 250-ml Erlenmeyer flasks.

Clean your buret thoroughly. Draw about 50 ml of the standard $KMnO_4$ solution from the carboy in the laboratory. Rinse the buret with a few milliliters of the $KMnO_4$ three times. Drain and then fill the buret with the $KMnO_4$ solution.

Prepare 150 ml of 1 M H_2SO_4 by pouring 25 ml of 6 M H_2SO_4 into 125 ml of H_2O, while stirring. Add 50 ml of this 1 M H_2SO_4 to *one* of the iron samples. The sample should dissolve completely. Without delay, titrate this iron solution with the $KMnO_4$ solution. When a light yellow color develops in the iron solution during the titration, add 3 ml of 85 per cent H_3PO_4. *CAUTION:* Caustic reagent. Continue the titration until you obtain the first pink color that persists for 15 to 30 seconds. Repeat the titration with the other two samples.

Name _____ Section _____

DATA AND CALCULATIONS: Determination of Iron by Permanganate Titration

Mass of sample tube plus contents _____ g

Mass after removing Sample I _____ g

Mass after removing Sample II _____ g

Mass after removing Sample III _____ g

Molarity of standard $KMnO_4$ solution _____ M

Sample	I	II	III
Initial buret reading	_____ ml	_____ ml	_____ ml
Final buret reading	_____ ml	_____ ml	_____ ml
Volume $KMnO_4$ required	_____ ml	_____ ml	_____ ml
Moles $KMnO_4$ required	_____	_____	_____
Moles Fe^{2+} in sample	_____	_____	_____
Mass of Fe in sample	_____ g	_____ g	_____ g
Mass of sample	_____ g	_____ g	_____ g
% Fe in sample	_____ %	_____ %	_____ %

Unknown no. _____

Name _____ **Section** _____

ADVANCE STUDY ASSIGNMENT: Determination of Iron by Permanganate Titration

1. Write the balanced net ionic equation for the reaction between MnO_4^- ion and Fe^{2+} ion in acid solution.

2. How many moles of Fe^{2+} ion can be oxidized by 1.2×10^{-2} moles MnO_4^- ion in the reaction in Question 1?

_____ moles

3. A solid sample containing some Fe^{2+} ion weighs 1.705 g. It requires 36.44 ml 0.0244 M $KMnO_4$ to titrate the Fe^{2+} in the dissolved sample to a pink end point.

   a. How many moles MnO_4^- ion are required?

_____ moles

   b. How many moles Fe^{2+} are there in the sample?

_____ moles

   c. How many grams of iron are there in the sample?

_____ g

   d. What is the per cent Fe in the sample?

_____ %

4. What is the per cent Fe in iron(II) ammonium sulfate hexahydrate, $Fe(NH_4)_2(SO_4)_2 \cdot 6\,H_2O$?

_____ %

EXPERIMENT

32 • Synthesis of Some Coordination Compounds

Some of the most interesting research in inorganic chemistry has involved the preparation and study of the properties of those substances known as coordination compounds. These compounds, sometimes called complexes, are typically salts that contain *complex ions*. A complex ion is an ion that contains a central metal ion to which are bonded small polar molecules or simple ions; the bonding in the complex ion is through coordinate covalent bonds, which ordinarily are relatively weak.

In this experiment we shall be concerned with the synthesis of two coordination compounds:

$$A. \ [Cu(NH_3)_4] \, SO_4 \cdot H_2O \qquad B. \ [Co(NH_3)_6] \, Cl_3$$

The complex ions in these substances are enclosed in brackets to indicate those species which are bonded to the central ion. In this experiment you will prepare the complex ions by making use of reactions in which substituting *ligands*, or coordinating species, replace other ligands on the central ion. The reactions will usually be carried out in water solution, in which the metallic cation will initially be present in the simple hydrated form; addition of a reagent containing a complexing ligand will result in an exchange reaction of the sort

$$Cu(H_2O)_4{}^{2+}(aq) + 4\,NH_3(aq) \rightleftharpoons Cu(NH_3)_4{}^{2+}(aq) + 4\,H_2O \qquad (1)$$

In many reactions involving complex ion formation the rate of reaction is very rapid, so that the thermodynamically stable form of the ion is the one produced. Such reactions obey the law of chemical equilibrium and can thus be readily controlled as to direction by a change in the reaction conditions. Reaction 1 proceeds readily to the right in the presence of NH_3 in moderate concentrations. However, by decreasing the NH_3 concentration, for example by the addition of acid to the system, we can easily regenerate the hydrated copper cation. Complex ions that undergo very fast exchange reactions, such as those in Reaction 1, are called *labile*.

Most but by no means all complex ions are labile. Some complex ions, including one to be studied in this experiment, will exchange ligands only slowly. For such species, called *inert* or *nonlabile*, the complex ion produced in a substitution reaction may be the one which is kinetically rather than thermodynamically favored. Alteration of reaction conditions, perhaps by the addition of a catalyst, may change the relative rates of formation of possible complex products and so change the complex ion produced in the reaction. In your experiment you will use a catalyst, activated charcoal, in the preparation of the complex ion in compound B. In the absence of the catalyst the complex ion that would form would be $Co(NH_3)_5H_2O^{3+}$. The Co(III) present in these species is stabilized in complex ions containing NH_3 ligands.

Many complex ions are highly colored, both in solution and in the solid salt. An easy way to determine whether a complex ion is labile is to note whether a color change occurs in a solution containing the ion when a good complexing ligand is added. You may wish to use this procedure in observing relative rates of substitution reactions involving the complex ions you prepare.

EXPERIMENTAL PROCEDURE

A. Preparation of $[Cu(NH_3)_4]\,SO_4 \cdot H_2O$

$$Cu(H_2O)_4{}^{2+}(aq) + SO_4{}^{2-}(aq) + 4\,NH_3(aq) \rightarrow [Cu(NH_3)_4]\,SO_4 \cdot H_2O(s) + 3\,H_2O$$

Weigh out 7.0 g of $CuSO_4 \cdot 5\,H_2O$ on a triple-beam or other rough balance. Weigh the solid either on a piece of paper or in a small beaker and not directly on the balance pan. Transfer the solid copper sulfate to a 125 ml Erlenmeyer flask and add 15 ml of water. Heat the flask to dissolve the solid, then cool to room temperature.

Carry out the remaining steps under the hood. Add 15 M NH_3 solution, a few ml at a time, swirling the flask to mix the reagents, until the first precipitate has completely dissolved. All the copper should now be present in solution as the complex ion $Cu(NH_3)_4{}^{2+}$.

The sulfate salt of this complex cation can be precipitated by the addition of a liquid, such as ethyl alcohol, in which $Cu(NH_3)_4SO_4 \cdot H_2O$ is insoluble. Add 10 ml of 95% ethyl alcohol to the solution; this should result in the formation of a deep blue precipitate of $Cu(NH_3)_4SO_4 \cdot H_2O$. Filter the solution through a Buchner funnel, using suction. Wash the solid in the funnel by adding two 5 ml portions of 95% ethanol. Dry the solid by pressing it between two pieces of filter paper. Put the crystals on a piece of weighed filter paper and let them dry further in the air. When they are thoroughly dry, weigh them on the paper to 0.1 g.

B. Synthesis of Hexamminecobalt(III) Chloride, $[Co(NH_3)_6]\,Cl_3$

$$2\,Co(H_2O)_6{}^{2+}(aq) + 6\,Cl^-(aq) + 12\,NH_3(aq) + H_2O_2(aq)$$
$$\rightarrow 2\,[Co(NH_3)_6]\,Cl_3(s) + 12\,H_2O + 2\,OH^-(aq)$$

Obtain a 5-ml sample of cobalt(II) chloride solution (containing 1.0 g $CoCl_2$). Pour this solution into a 150-ml beaker and add 1.2 g NH_4Cl, 0.1 g activated charcoal (Norite), and 5 ml distilled water. Heat the solution to the boiling point, stirring to complete the solution of the NH_4Cl.

Cool the beaker in an ice bath, and slowly add 5 ml 15 M NH_3. Stir until well mixed and then slowly, drop by drop, with stirring, add 5 ml of 10% hydrogen peroxide, H_2O_2.

When the bubbling has stopped, place the 150-ml beaker in a 600-ml beaker containing 100 ml water at about 60°C. Leave the beaker in the water bath for 20 minutes, holding the temperature of the bath at 60 ± 5°C by judicious use of your Bunsen burner. Stir the mixture occasionally.

Remove the 150-ml beaker from the water bath and return it to the ice bath. Cool for about 5 minutes, stirring to promote crystallization of the crude product. Set up a Büchner funnel, and, with suction, filter the mixture through the filter paper in the funnel. Discard the filtrate.

Scrape the solid from the filter paper into a 150-ml beaker. Add 50 ml H_2O and 2 ml 12 M HCl. Heat the mixture to boiling, with stirring. Using distilled water, rinse out the suction flask used with the Büchner funnel. Holding the beaker with a pair of tongs, filter the hot mixture through the funnel, with suction on. The charcoal is removed at this point and should remain on the filter paper; the filtrate should be a golden yellow.

Transfer the filtrate to a 150-ml beaker and add 8 ml 12 M HCl. Place the beaker in an ice bath and stir for several minutes to promote formation of the golden crystals of $[Co(NH_3)_6]\,Cl_3$. Separate the crystals by filtration through the Büchner funnel. Turn the suction off, add 10 ml 95% ethanol, and let stand for about 10 seconds. Turn the suction on and pull off the ethanol, which should carry with it much of the water and HCl remaining on the crystals. Draw air through the crystals for several minutes. Transfer the product to a piece of weighed filter paper. Let it dry for a few minutes and weigh it on the paper to 0.1 g.

C. Relative Lability of Complex Ions. Dissolve a small portion (about 0.5 g) of the complexes from Parts A and B, $[Cu(NH_3)_4] SO_4 \cdot H_2O$ and $[Co(NH_3)_6] Cl_3$, in a few ml of water. Note the color and observe the effect of the addition of a few drops of 12 M HCl.

On completion of Part C, show your products from Parts A and B to your instructor for evaluation.

CAUTION: In these experiments you will be using 15 M NH_3, 12 M HCl, and a 10% solution of H_2O_2. These are caustic reagents, so use them carefully. Avoid getting them on your skin or clothing, and don't breathe the vapors of NH_3 or HCl. If you do come into contact with any of these reagents, wash them off thoroughly with water.

DATA AND OBSERVATIONS: Synthesis of Some Coordination Compounds

	$[Cu(NH_3)_4] SO_4 \cdot H_2O$	$[Co(NH_3)_6] Cl_3$
Mass of filter paper	_____ g	_____ g
Mass of filter paper plus product	_____ g	_____ g
Mass of product	_____ g	_____ g
Theoretical yield	_____ g	_____ g
Percentage yield	_____ %	_____ %

Observations Regarding Lability of Complexes (Part C)

	Color of solid	Color in H_2O solution	Color on addition of HCl
$Cu(NH_3)_4{}^{2+}$	_____	_____	_____
$Co(NH_3)_6{}^{3+}$	_____	_____	_____

Conclusions

What evidence, if any, do you have that the formulas of the two compounds you prepared are correct, either as to the nature of the atoms or groups present or as to the number of the groups present? Consider the nature of the possible cations, anions, and ligands in the compounds.

ADVANCE STUDY ASSIGNMENT: Synthesis of Some Coordination Compounds

1. Calculate the theoretical yields of the compounds to be prepared in this experiment. The metal ion in both cases is the limiting reagent. Hint: Find the number of moles of Cu(II) in the sample of $CuSO_4 \cdot 5\ H_2O$ that you used. That will equal the number of moles of $[Cu(NH_3)_4]\ SO_4 \cdot H_2O$ that could theoretically be prepared. Proceed in a similar way for the synthesis involving Co(II).

A.

_____g

B.

_____g

2. How could you establish that the metal ion is the limiting reagent?

3. How would you formally name the compound $[Cu(NH_3)_4]SO_4 \cdot H_2O$ prepared in Part A?

4. Give the formulas of the following compounds:

    a. Dichlorotetramminechromium(III) nitrate

    b. Dichlorodiaquoplatinum(II)

EXPERIMENT

33 • Preparation of Aspirin

One of the simpler organic reactions that one can carry out is the formation of an ester from an acid and an alcohol:

$$\underset{\text{an acid}}{R-\overset{\displaystyle O}{\overset{\|}{C}}-OH} \;+\; \underset{\text{an alcohol}}{HO-R'} \;\rightarrow\; \underset{\text{an ester}}{R-\overset{\displaystyle O}{\overset{\|}{C}}-O-R'} \;+\; H_2O \qquad (1)$$

In the equation, R and R′ are H atoms or organic fragments like CH_3, C_2H_5, or more complex aromatic groups. There are many esters, since there are many organic acids and alcohols, but they all can be formed, in principle at least, by Reaction 1. The driving force for the reaction is in general not very great, so that one ends up with an equilibrium mixture of ester, water, acid, and alcohol.

There are some esters which are solids because of their high molecular weight or other properties. Most of these esters are not soluble in water, so they can be separated from the mixture by crystallization. This experiment deals with an ester of this sort, the substance commonly called aspirin. Aspirin is the active component in headache pills and is one of the most effective, relatively nontoxic, pain killers.

Aspirin can be made by the reaction of the —OH group in the salicyclic acid molecule with the carboxyl (—COOH) group in acetic acid:

$$\underset{\text{acetic acid}}{CH_3\overset{\displaystyle O}{\overset{\|}{C}}-OH} \;+\; \underset{\text{salicylic acid}}{HO-} \rightleftharpoons \underset{\text{aspirin}}{CH_3-\overset{\displaystyle O}{\overset{\|}{C}}-O-} \;+\; H_2O \qquad (2)$$

A better preparative method, which we will use in this experiment, employs acetic anhydride in the reaction instead of acetic acid. The anhydride can be considered to be the product of a reaction in which two acetic acid molecules combine, with the elimination of a molecule of water. The anhydride will react with the water produced in the esterification reaction and will tend to drive the reaction to the right. A catalyst, normally sulfuric or phosphoric acid, is also used to speed up the reaction.

acetic anhydride   salicylic acid   aspirin   acetic acid

The aspirin you will prepare in this experiment is relatively impure and should certainly not be taken internally, even if the experiment gives you a bad headache.

There are several ways by which the purity of your aspirin can be estimated. Probably the simplest way is to measure its melting point. If the aspirin is pure, it will melt sharply at the literature value of the melting point. If it is impure, the melting point will be lower than the literature value by an amount that is roughly proportional to the amount of impurity present.

A more quantitative measure of the purity of your aspirin sample can be obtained by determining the per cent salicylic acid it contains. Salicylic acid is the most likely impurity in the sample since, unlike acetic acid, it is not very soluble in water. Salicylic acid forms a highly colored magenta complex with Fe(III). By measuring the absorption of light by a solution containing a known amount of aspirin in excess Fe^{3+} ion, one can easily determine the per cent salicylic acid present in the aspirin.

EXPERIMENTAL PROCEDURE   WEAR YOUR SAFETY GLASSES WHILE PERFORMING THIS EXPERIMENT

Weigh a 50 ml Erlenmeyer flask on a triple beam or top loading balance and add 2.0 g of salicylic acid. Measure out 5.0 ml of acetic anhydride in your graduated cylinder, and pour it into the flask in such a way as to wash any crystals of salicylic acid on the walls down to the bottom. Add 5 drops of 85 per cent phosphoric acid to serve as a catalyst. *Both acetic anhydride and phosphoric acid are reactive chemicals which can give you a bad chemical burn, so use due caution in handling them.* If you get any of either on your hands or clothes, wash thoroughly with soap and water.

Clamp the flask in place in a beaker of water supported on a wire gauze on a ring stand. Heat the water with a Bunsen burner to about 75°C, stirring the liquid in the flask occasionally with a stirring rod. Maintain this temperature for about 15 minutes, by which time the reaction should be complete. *Cautiously,* add 2 ml of water to the flask to decompose any excess acetic anhydride. There will be some hot acetic acid vapor evolved as a result of the decomposition.

When the liquid has stopped giving off vapors, remove the flask from the water bath and add 20 ml of water. Let the flask cool for a few minutes in air, during which time crystals of aspirin should begin to form. Put the flask in an ice bath to hasten crystallization and increase the yield of product. If crystals are slow to appear, it may be helpful to scratch the inside of the flask with a stirring rod. Leave the flask in the ice bath for at least 5 minutes.

Collect the aspirin by filtering the cold liquid through a Buchner funnel using suction. Turn off the suction and pour about 5 ml of ice-cold distilled water over the crystals; after about 15 seconds turn on the suction to remove the wash liquid along with most of the impurities. Repeat the washing process with another 5 ml sample of ice-cold water. Draw air through the funnel for a few minutes to help dry the crystals and then transfer them to a piece of dry, weighed, filter paper. Weigh the sample on the paper to ± 0.1 g.

Test the solubility properties of the aspirin by taking samples of the solid the size of a pea on your spatula and putting them in separate 1 ml samples of each of the following solvents and stirring:

1. Toluene, $C_6H_5CH_3$, nonpolar aromatic
2. Hexane, C_6H_{14}, nonpolar aliphatic
3. Ethyl acetate, $C_2H_5OCOCH_3$, aliphatic ester
4. Ethyl alcohol, C_2H_5OH, polar aliphatic, hydrogen bonding
5. Acetone, CH_3COCH_3, polar aliphatic, nonhydrogen bonding
6. Water, highly polar, hydrogen bonding

To determine the melting point of the aspirin, add a small amount of your prepared sample to a melting point tube (made from 5-mm tubing), as directed by your instructor. Shake the solid down by tapping the tube on the bench top, using enough solid to give you a depth of about 5 mm. Set up the apparatus shown in Figure 33.1. Fasten the melting point tube to the thermometer with a small rubber band, which should be above the surface of the oil. The thermometer bulb and sample should be about 2 cm above the bottom of the tube. Heat the oil bath *gently*, especially

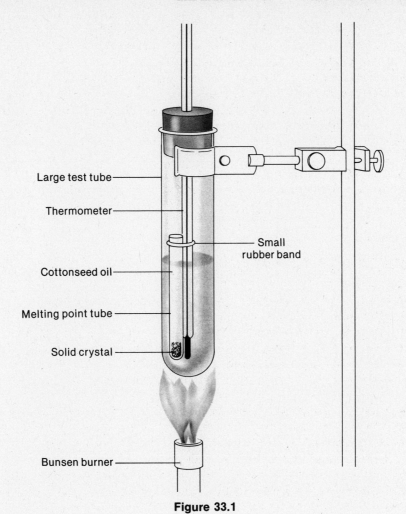

Large test tube

Thermometer

Small
rubber band

Cottonseed oil

Melting point tube

Solid crystal

Bunsen burner

Figure 33.1

after the temperature gets above 100°C. As the melting point is approached, the crystals will begin to soften. Report the melting point as the temperature at which the last crystals disappear.

To analyze your aspirin for its salicylic acid impurity, weigh out 0.10 ± 0.01 g of your sample into a weighed 100-ml beaker. Dissolve the solid in 5 ml 95% ethanol. Add 5 ml 0.025 M $Fe(NO_3)_3$ in 0.5 M HCl and 40 ml distilled water. Make all volume measurements with a graduated cylinder. Stir the solution to mix all reagents.

Rinse out a spectrophotometer tube with a few ml of the solution and then fill the tube with that solution. Measure the absorbance of the solution at 525 nm. The absorbance measurement should be made *within 5 minutes* of the time the sample was dissolved in the ethanol, since aspirin will gradually decompose in solution, producing salicylic acid and acetic acid. From the calibration curve or equation provided, calculate the per cent salicylic acid in the aspirin sample.

Name _____ **Section** _____

DATA AND RESULTS: Preparation of Aspirin

Mass of salicylic acid used   _____ g

Volume of acetic anhydride used   _____ ml

Mass of acetic anhydride used
(density = 1.08 g/ml)

                                                            _____ g

Mass of aspirin obtained   _____ g

Theoretical yield of aspirin

                                                            _____ g

Percentage yield of aspirin

                                                            _____ %

Melting point of aspirin   _____ °C

Absorbance of aspirin solution   _____

Per cent salicylic acid impurity   _____ %

Solubility properties of aspirin

   Toluene   _____   Ethyl alcohol _____

   Hexane   _____   Acetone   _____

   Ethyl acetate _____   Water   _____

S = soluble   I = insoluble   SS = slightly soluble

Underline those characteristics listed below which would be likely to be present in a good solvent for aspirin.

   organic   aliphatic   polar   hydrogen bonding

   inorganic   aromatic   nonpolar   nonhydrogen bonding

ADVANCE STUDY ASSIGNMENT:　Preparation of Aspirin

1.　Calculate the theoretical yield of aspirin to be obtained in this experiment, starting with 2.0 g of salicylic acid and 5.0 ml of acetic anhydride (density = 1.08 g/ml).

————————— g

2.　If 1.9 g of aspirin were obtained in this experiment, what would be the percentage yield?

————————— %

3.　The name acetic anhydride implies that the compound will react with water to form acetic acid. Write the equation for the reaction.

4.　Identify R and R′ in Equation 1 when the ester, aspirin, is made from salicylic acid and acetic acid.

5.　Write the equation for the reaction by which aspirin decomposes in an aqueous ethanol solution.

EXPERIMENT

34 • Preparation of a Synthetic Resin

In this experiment we will prepare and examine the properties of one of the most common polymers, polystyrene, made from styrene, C_6H_5—CH=CH_2. Styrene is relatively easy to polymerize; the plastic made from it is typically quite hard and transparent. In pure polystyrene, the chain is unbranched:

styrene   section of a polystyrene molecule

Polymers containing unbranched chains are thermoplastic, which means that they can be melted and then cast or extruded into various shapes. They also can usually be dissolved in some organic solvents, forming viscous liquids. If a polymer is cross-linked so that its chains are bonded together at regular or random positions, it will usually neither melt nor dissolve readily; such a material is called thermosetting and is usually polymerized in a mold in the shape of the article desired.

Styrene will polymerize to a crystalline solid if you simply heat it. The polymerization reaction itself evolves heat, however, and once the reaction gets started it tends to increase in rate and can get out of control; the simplest commercial process polymerizes styrene this way, and one of the important problems is to provide adequate cooling as the reaction proceeds.

We will polymerize styrene under somewhat different conditions, using an emulsion polymerization, in which the styrene is dispersed into droplets in water. In this process, temperature control is easy, and the polymer is produced in the form of easy-to-handle beads. By carrying out the reaction in the presence of divinylbenzene, which can react at two double-bonded positions, we will make a cross-linked polymer very similar in structure to that of an ion-exchange resin. Divinylbenzene is much like styrene except that there are two ethylene groups rather than one attached to each benzene ring. The compound has three isomers:

paradivinylbenzene   metadivinylbenzene   orthodivinylbenzene

The commercially available divinylbenzene which we use contains mainly the *para* and *meta* isomers, in about equal amounts.

The resin produced will have a structure similar to that indicated below:

Relatively few divinylbenzene molecules are required for the cross-linking. The material produced is really a copolymer of styrene and divinylbenzene.

After preparing the polymer, we will compare its melting point and solubility properties with those of linear polystyrene.

EXPERIMENTAL PROCEDURE

WEAR YOUR SAFETY GLASSES WHILE
PERFORMING THIS EXPERIMENT

Add 1.0 g gelatin to 100 ml of distilled water in a 250-ml Erlenmeyer flask. Heat the water to near boiling, while occasionally stirring the gelatin to disperse it. When the water gets hot, the gelatin will swell and gradually go into solution. There will be some tendency to foam up, and when this begins to occur, stop heating and stir. (If you have to, add a little water from your wash bottle to control the foaming.) With careful heating and stirring, get all of the gelatin to dissolve. Then add about 0.05 g of calcium phosphate, $Ca_{10}(OH)_2(PO_4)_6$, to the solution and stir. The solid will not dissolve but will disperse uniformly in the liquid.

Fill a 1000-ml beaker about half-full of water and put the flask in the beaker as in Figure 34.1. The water level in the beaker should be a few centimeters above that in the flask. Heat the water in the beaker to about 90°C.

While the water is heating, measure out 8 ml of styrene and 3 ml of divinylbenzene into a 30-ml beaker. Your instructor will add 100 mg benzoyl peroxide (very reactive!) to the beaker. Swirl the beaker to dissolve the solid and initiate the reaction; keep the beaker in ice water while waiting for the water to heat.

When the water in the large beaker is at about 90°C, remove the flask from the water bath, and slowly, with swirling, add the cold styrene to the gelatin solution. Continue to swirl the mixture for about 20 seconds. Stopper the flask, loosely, with a cork to minimize vaporization of the styrene, and put the flask back in the water bath.

The styrene–divinylbenzene mixture will float on the liquid as small droplets. The purpose of the gelatin and calcium phosphate is to stabilize this dispersion as polymerization proceeds. Heat the polymerizing mixture for about an hour, keeping the temperature of the water bath between 85° and 95°C. Stir the mixture frequently with your stirring rod to keep the droplets dispersed, keeping the flask covered except while stirring.

Initially the droplets will be liquid, and the color will be somewhat yellow. The color will gradually disappear. When it is about gone, in 10–15 minutes, the droplets will have become partially polymerized, a bit sticky, and may tend to agglomerate into chunks. If you observe the droplets carefully, and stir frequently, you can prevent this happening, at least for the most part. Once that stage is passed, it will be easier to keep the droplets dispersed. If all goes well, by the

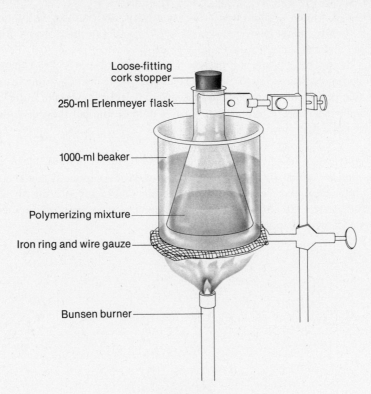

Loose-fitting cork stopper

250-ml Erlenmeyer flask

1000-ml beaker

Polymerizing mixture

Iron ring and wire gauze

Bunsen burner

Figure 34.1

end of the hour the polymerization will be complete, and the polymer will consist mainly of small beads, varying in size from very small up to about 2 mm in diameter, with perhaps a few larger chunks.

When the heating period is over, remove the flask from the water bath, and pour its contents into 300 ml of distilled water in a 600-ml beaker. Stir to dilute the gelatin, and then filter the mixture through a Buchner funnel, using suction, and filtering through a piece of filter paper. Wash the polymer with several portions of distilled water. Turn off the suction and transfer the polymer particles to a piece of paper for further drying. When they are dry, weigh them.

The polymer you have prepared differs from ordinary polystyrene in several ways, and in the last part of the experiment, we will compare some of its properties to those of a typical polystyrene.

Tear some polystyrene from a foam coffee cup into pieces small enough to fit into a small test tube. Fill three such tubes with those pieces. The polymer in the foam pieces is essentially linear polystyrene with few branches and no cross-links. Put 2 ml toluene in one of the test tubes and 2 ml acetone in the second; shake to get the foam wet with solvent.

toluene

acetone

Heat the third tube gently in the Bunsen flame, noting whether the polymer melts or decomposes. Estimate the temperature at which the change occurs, but don't try to measure it. Poke the material with your stirring rod to aid in establishing its viscosity. Shake the tubes containing solvent and note whether the polymer has dissolved. If it hasn't, put the tubes in a hot water bath (use a 250-ml beaker), to speed up the solution process. Do not boil the solvent, however. Note the properties of the final mixture, particularly the viscosity and clarity of any solutions that form.

If you are able to obtain a solution with either solvent, pour a drop of solution on to some water in a 600-ml beaker. Blow gently on the surface to evaporate the solvent. Pick up the film with a stirring rod and note its thickness and strength.

Put a few of the polymer beads you prepared in small test tubes and test them as before for solubility in toluene and acetone. Try to melt the beads, by gentle heating. Compare their behavior on heating with that of the polystyrene foam. If your polymerization was not quite complete, your beads may be somewhat rubbery, rather than hard. Compare the hardness of your beads before heating with that after they have been heated.

CAUTION: In this experiment, the laboratory should be well ventilated, since we use several volatile organic liquids. Avoid contact with the liquids and their vapors. If you keep the polymerizing mixture lightly stoppered, very little styrene vapor should get into the air. Open the flask momentarily while you stir the mixture. If you get some liquid styrene or divinylbenzene on your hands, wash them off with soap and water.

DATA AND OBSERVATIONS: Preparation of a Synthetic Resin

Mass of styrene (density = 0.90 g/ml)   _____g

Mass of divinylbenzene (density = 0.90 g/ml)   _____g

Mass of resin   _____g

Theoretical yield   _____g

Percentage yield   _____%

Properties of polystyrene and prepared resin

	Polystyrene (foam cup)	Prepared resin
Behavior on heating	_____	_____
	_____	_____
Solubility in toluene	_____	_____
Solubility in acetone	_____	_____

How do you explain the difference in solubility of polystyrene in the two solvents?

Comment on the effects of cross-linking on the properties of polystyrene. Explain any change of hardness in your polymer that occurred when you heated it.

EXP. 34 Name ______________________ Section ______________

ADVANCE STUDY ASSIGNMENT: Preparation of a Synthetic Resin

1. Polyvinyl chloride is made by addition polymerization of vinyl chloride, $CH_2{=}\underset{\displaystyle Cl}{\underset{|}{C}}{-}H$. Sketch a section of the polyvinyl chloride molecule.

2. How much polyvinyl chloride could theoretically be made from 100 g of vinyl chloride?

______________ g

3. What per cent by mass of polyvinyl chloride is carbon?

______________ %

4. Polystyrene foam such as that used in coffee cups is made by a procedure very analogous to that used in this experiment. Can you suggest how the foamable polymer might be made and how it would be converted to the form of a foam coffee cup? Before answering, take a good look at a foam coffee cup.

5. Cross-linked polymers are much less soluble in organic solvents than linear polymers. Explain why that is the case.

EXPERIMENT

35 • Spot Tests for Some Common Ions

There are two broad categories of problems in analytical chemistry. Quantitative analysis deals with the determination of the amounts of certain species present in a sample; there are several experiments in this manual involving quantitative analysis, and you have probably performed some of them. The other area of analysis, called qualitative analysis, has a more limited purpose, namely, establishing whether given species are or are not present in detectable amounts in a sample. The remaining experiments in this manual will deal mainly with problems in qualitative analysis.

One can carry out the qualitative analysis of a sample in various ways. Probably the simplest approach, which we will use in this experiment, is to test for the presence of each possible component by adding a reagent which will cause the component, if it is in the sample, to react in a characteristic way. This method involves a series of "spot" tests, one for each component, carried out on separate samples of the unknown. The difficulty with this way of doing qualitative analysis is that frequently, particularly in complex mixtures, one species may interfere with the analytical test for another. Although interferences are common, there are many ions which can, under optimum conditions at least, be identified in mixtures by simple spot tests.

In this experiment we will use spot tests for the analysis of a mixture which may contain the following commonly encountered ions in solution:

$$CO_3^{2-} \qquad PO_4^{3-} \qquad Cl^- \qquad SCN^-$$

$$SO_4^{2-} \qquad CrO_4^{2-} \qquad C_2H_3O_2^- \qquad NH_4^+$$

The procedures we will use involve simple acid-base, precipitation, complex ion formation, or oxidation-reduction reactions. In each case you should try to recognize the kind of reaction that occurs, so that you can write the net ionic equation that describes it.

EXPERIMENTAL PROCEDURE

WEAR YOUR SAFETY GLASSES WHILE PERFORMING THIS EXPERIMENT

Carry out the test for each of the anions as directed. Repeat each test using a solution made by diluting the anion solution 9:1 with distilled water; use your 10 ml graduated cylinder to make the dilution and make sure you mix well before taking the sample for analysis. In some of the tests, a boiling water bath containing about 100 ml water in a 150 ml beaker will be needed, so set that up before proceeding. When performing a test, if no reaction is immediately apparent, stir the mixture with your stirring rod to mix the reagents. These tests can easily be used to detect the anions at concentrations of 0.02 M or greater, but in dilute solutions, careful observation may be required.

Test for the Presence of Carbonate Ion, CO_3^{2-}. Cautiously add 1 ml of 6 M HCl to 1 ml of 1 M Na_2CO_3 in a small test tube. With concentrated solutions, bubbles of carbon dioxide gas are immediately evolved. With dilute solutions, the effervescence will be much less obvious. Warming in the water bath, with stirring, will increase the amount of bubble formation. Carbon dioxide is colorless and odorless.

Test for the Presence of Sulfate Ion, SO_4^{2-}. Add 1 ml of 6 M HCl to 1 ml of 0.5 M Na_2SO_4. Add a few drops of 1 M $BaCl_2$. A white, finely divided precipitate of $BaSO_4$ indicates the presence of SO_4^{2-} ion.

Test for the Presence of Phosphate Ion, PO_4^{3-}.   Add 1 ml 6 M HNO_3 to 1 ml of 0.5 M Na_2HPO_4. Then add 1 ml of 0.5 M $(NH_4)_2MoO_4$ and stir thoroughly. A yellow precipitate of ammonium phosphomolybdate, $(NH_4)_3PO_4 \cdot 12\,MoO_3$, establishes the presence of phosphate. The precipitate may form slowly, particularly in the more dilute solution; if it does not appear promptly, put the test tube in the boiling water bath for a few minutes.

Test for the Presence of Chromate Ion, CrO_4^{2-}.   Chromate-containing solutions are yellow when neutral or basic and orange under acidic conditions. Chromate ion under highly acidic conditions is a good oxidizing agent and may oxidize such species as SCN^- or NH_4^+. Add 2 ml of 6 M HNO_3 to 1 ml of 0.5 M K_2CrO_4. If reduction of chromate occurs at this point due to the presence of interfering species, a color change to pale blue will be observed shortly after addition of the acid and may be taken as proof of the presence of chromate ion. If no color change is observed, cool the test tube under the water tap and add 1 ml of 3 per cent H_2O_2. If chromate ion is present, a rapidly fading blue color caused by the formation of unstable CrO_5 will be observed.

Test for the Presence of Thiocyanate Ion, SCN^-.   Add 1 ml 6 M acetic acid to 1 ml of 0.5 M KSCN and stir. Add one or two drops of 0.1 M $Fe(NO_3)_3$. A deep red coloration due to formation of $FeSCN^{2+}$ complex ion is proof of the presence of SCN^- ion.

Test for the Presence of Chloride Ion, Cl^-.   Add 1 ml of 6 M HNO_3 to 1 ml of 0.5 M NaCl. Add 2–3 drops of 0.1 M $AgNO_3$. A white precipitate of AgCl will form if chloride ion is present.

If thiocyanate ion is present, it will interfere with this test, since it will also form a white precipitate. If the sample contains SCN^- ion, put 1 ml of the solution in a small 30 or 50 ml beaker and add 1 ml of 6 M HNO_3. Boil the solution gently until the volume is decreased to one-half its original value; this will oxidize the thiocyanate and remove the interference. Then add another ml of 6 M HNO_3 and a few drops of $AgNO_3$ solution, forming white AgCl in the presence of Cl^- ion as before.

Test for the Presence of Acetate Ion, $C_2H_3O_2^-$.   Add 1 ml 3 M H_2SO_4 to 1 ml of 2 M $NaC_2H_3O_2$ and stir. If acetate ion is present, acetic acid, with the characteristic odor of vinegar, is formed. The intensity of the odor is enhanced if the tube is warmed for 30 seconds in the boiling water bath. If the acetate ion concentration is low, boil the neutral solution carefully to near dryness before acidifying it.

Test for the Presence of Ammonium Ion, NH_4^+.   Add 1 ml of 6 M NaOH to 1 ml of 0.5 M NH_4Cl. If NH_4^+ ion is present, NH_3 will form and can be detected by its characteristic odor. For a more sensitive test, pour the 1 ml of sample into a small beaker; moisten a piece of red litmus paper and put it on the bottom of a watch glass; cover the beaker with the watch glass and gently heat the liquid to the boiling point. Do not boil it and be careful that no liquid comes in contact with the litmus paper. If NH_4^+ ion is present, the litmus paper will gradually turn blue as it is exposed to the vapors of NH_3 that will evolve. Remove the watch glass and try to smell the ammonia.

When you have completed all of the tests, obtain an unknown from your laboratory supervisor and analyze it by applying the tests to separate 1 ml portions. The unknown will contain 3 or 4 of the ions on the list, so your test for a given ion may be affected by the presence of others. For the most part, the properties of the mixture will be a superposition of the properties of its components, but it is possible that under some conditions the ions will react in unexpected ways. Where a test does not go quite according to the rules, try to deduce why the sample may have behaved as it did. When you think you have properly analyzed your unknown, you may, if you wish, make a "known" which has the composition you found and test it to see if it has the properties of your unknown.

Name _____ Section _____

OBSERVATIONS AND REPORT SHEET: Spot Tests for Some Common Ions

Observations and Comments on Spot Tests

Ion	Stock Solution	9:1 Dilution	Unknown

CO_3^{2-}

_____

SO_4^{2-}

_____

PO_4^{3-}

_____

CrO_4^{2-}

_____

Cl^-

_____

$C_2H_3O_2^-$

_____

SCN^-

_____

NH_4^+

_____

Unknown No. _____ contains _____

ADVANCE STUDY ASSIGNMENT: Spot Tests for Some Common Ions

1. Each of the observations listed below was made on a different solution. Given the observation, state which ion studied in this experiment is present. If the test is not definitive, indicate that with a question mark, "?".

    a. Addition of 6 M NaOH to the solution produces a vapor with a characteristic odor.
       Ion present:

    b. Addition of 6 M HNO_3 produces a color change.
       Ion present:

    c. Addition of 6 M HNO_3 produces an effervescence.
       Ion present:

    d. Addition of 6 M HNO_3 plus 0.1 M $AgNO_3$ produces a precipitate.
       Ion present:

    e. Addition of 6 M HNO_3 plus 1 M $BaCl_2$ produces a precipitate.

    f. Addition of 6 M H_2SO_4 produces a vapor with a characteristic odor.
       Ion present:

    g. Addition of 6 M HNO_3 plus 0.5 M $(NH_4)_2MoO_4$ produces a precipitate.

2. An unknown containing one or more of the ions studied in this experiment has the following properties:
    a. No effect on addition of 6 M HNO_3.
    b. No effect on addition of 0.1 M $AgNO_3$ to above solution.
    c. White precipitate on addition of 1 M $BaCl_2$ to solution in a.
    d. Yellow precipitate on addition of $(NH_4)_2MoO_4$ to solution in a.
On the basis of this information which ions are present, which are absent, and which are in doubt?

       Present   Absent   In doubt

3. The chemical reactions that are used in the anion spot tests in this experiment are for the most part simple precipitation or acid-base reactions. Given the information in each test procedure, try to write the net ionic equation for the key reaction in each test.

    a. CO_3^{2-}

    b. SO_4^{2-}

Continued on following page

c. PO_4^{3-} (Reactants are HPO_4^{2-}, NH_4^+, MoO_4^{2-}, and H^+; products are $(NH_4)_3PO_4 \cdot 12\ MoO_3$ and H_2O; no oxidation or reduction occurs)

d. CrO_4^{2-} (Reactants are $Cr_2O_7^{2-}$, H_2O_2, and H^+; products are CrO_5 and H_2O)

e. SCN^-

f. Cl^-

g. $C_2H_3O_2^-$

h. NH_4^+

EXPERIMENT
36 • Identification of Unknown Solids

In the previous experiment you worked with some of the tests that are used to establish the presence of some of the most common anions in a solution. These tests can, in general, be applied to any solution in which the anions may be present and to any solid that is soluble in water.

In this experiment you will be asked to identify, on the basis of their chemical or physical properties, some solid ionic compounds. We have prepared a group of containers, each of which contains a single pure ionic substance. For each container there are two formulas, one of which is that of the substance present in it. Your problem is simply to establish which of the formulas is correct. Your grade will depend on the the number of substances you can identify in the time allowed.

Many of the substances can be identified on the basis of their anions, so that you can use the procedures in Experiment 34 to find the correct formula. In other cases the compounds may both be insoluble or may both have the same anion, so that other approaches will be necessary. One very useful procedure is to consider the solubilities of the two compounds in different reagents or the solubilities of products which can be made from those substances.

Probably the most important solvent to consider is water. One of the compounds may be water-soluble and the other may not be. The answer to such a question may possibly be apparent to you from your previous work in chemistry. If it is not, then note the information given in Appendix II, where the solubility properties of about 200 of the most common inorganic compounds are summarized. Properly used, Appendix II will allow you to make a great many decisions regarding solubilities of ionic substances under different conditions. For example, if you wished to distinguish $ZnSO_4$ from $BaSO_4$, you would find that in the row for Zn^{2+} and the column for SO_4^{2-} there is an S which, according to the key, means that the compound, $ZnSO_4$ in this case, is soluble in water (up to at least 0.1 M). In the row for Ba^{2+} and the column for SO_4^{2-} there is an I, which means that the compound $BaSO_4$ is insoluble in all common reagents. Thus, to distinguish $ZnSO_4$ from $BaSO_4$, we need merely to add water.

Differences in solubility in solvents other than water can often be used to good advantage. Although most carbonates, phosphates, chromates, hydroxides, and oxides are insoluble in water, they are soluble in 6 M solutions of strong acids. In Appendix II this is indicated by an A. This property would allow you to distinguish $BaSO_4$ from $BaCO_3$, or CuS from CuO. Another useful distinguishing solvent is 6 M NH_3, which dissolves many insoluble compounds by forming complex ions with the cation in that compound; in Appendix II this is indicated by an N. Given a sample that might be $Mg(OH)_2$ or $Zn(OH)_2$, both water-insoluble, and both soluble in strong acids, one could dissolve $Zn(OH)_2$ in 6 M NH_3, as indicated by N at the Zn^{2+}, OH^- intersection. Mg^{2+} does not form an ammonia complex ion, no N at the Mg^{2+}, OH^- intersection, so $Mg(OH)_2$ would not dissolve in 6 M NH_3.

If both compounds are soluble in water or another common solvent and contain the same anion, then a precipitating reagent that reacts with one of the cations can be very helpful. Say, for example, you needed to distinguish between $MgCl_2$ and $CdCl_2$, both of which are water-soluble. From examination of Appendix II you could determine that $MgCrO_4$ is water-soluble while $CdCrO_4$ is not. Addition of 1 M K_2CrO_4 to the water solution of the unknown should allow you to identify the substance present. In a more complicated case, with both compounds insoluble in water but soluble in a common solvent, it might be possible to find a selective precipitating reagent. If the unknown might be CuS or PbS, considerable examination of Appendix II would reveal that, although both substances are very insoluble, they both dissolve in hot 6 M HNO_3. Following treatment of the solid, which would put Cu^{2+} or Pb^{2+} into the solution, one could

TABLE 36.1 OUTLINE OF GENERAL PROCEDURE FOR IDENTIFYING THE SOLID PRESENT IN UNKNOWN

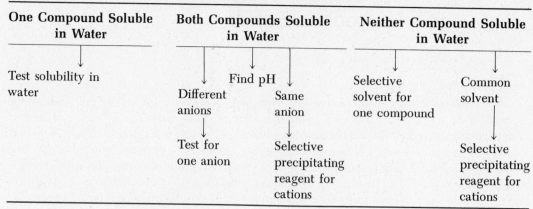

One Compound Soluble in Water	Both Compounds Soluble in Water		Neither Compound Soluble in Water	
Test solubility in water	Find pH		Selective solvent for one compound	Common solvent
	Different anions	Same anion		
	Test for one anion	Selective precipitating reagent for cations		Selective precipitating reagent for cations

identify the ion by its color, since Cu^{2+} is blue, or by precipitation on addition of 3 M H_2SO_4, which Appendix II tells us would bring down $PbSO_4$ but not $CuSO_4$.

Sometimes acidic or basic properties can be applied very easily to distinguish compounds. Anions like CO_3^{2-} or OH^- will always make their solutions basic. HSO_4^- ion is highly acidic. Color can sometimes be employed, as with the Cu^{2+} in the previous example. In Appendix II we have indicated the colors of the cations and anions that possess distinctive colors.

Although oxidation-reduction reactions are not often needed for distinguishing between compounds, they are sometimes very useful, as is the case with CuS and PbS. Halide ions in solution can be most easily distinguished by treatment with chlorine water, which will oxidize Br^- to Br_2 and I^- to I_2, both of which can be extracted into hexane, in which they have characteristic colors (recall Experiment 11).

Although it is not feasible to present a procedure that will work for every possible pair of compounds, we have presented in Table 36.1 a general outline which should be effective for dealing with most pairs. It essentially summarizes much of the material presented earlier in this discussion.

EXPERIMENTAL PROCEDURE   WEAR YOUR SAFETY GLASSES WHILE PERFORMING THIS EXPERIMENT

On each laboratory table there will be a set of numbered bottles, each bottle containing a pure solid compound. With each set there is a list on which you will find the number of each bottle and the formulas of two substances the bottle may contain.

When your instructor tells you to begin, select a bottle from the set and write your name on the line alongside the bottle number and its possible contents. Carry out tests to determine which of the two compounds the bottle actually contains. When you are able to decide, write the formula of the compound present on the data sheet, along with the bottle number and the two compounds between which you had to distinguish. On the same line write the net ionic equation for the reaction by which you were able to identify the compound. Go to your instructor, who will tell you if your reasoning is correct. Then proceed to select another bottle and determine its contents. Do as many analyses as you can in the time allowed.

At the end of about half an hour, you will be given another set of unknowns on which to work. You will be allowed to work on a total of four sets of bottles, and you will have about half an hour to spend on each set. For each of the first seven analyses you do correctly, you will obtain one point. For each correct analysis above seven, you will get half a point.

Only one student may work with a given unknown in a given set. A given reaction may be used only *once* for distinguishing between any two compounds; if you use the reaction $Ag^+ + Cl^- \rightarrow AgCl(s)$ to identify a substance, you may not use that reaction again to distinguish between another pair of possibilities.

Name _____ Section _____

REPORT SHEET: Identification of Unknown Solids

Unknown Number	Possible Compounds	Compound Present	Identifying reaction (Net ionic equation)
_____	_____	_____	_____
_____	_____	_____	_____
	_____		
_____	_____	_____	_____
	_____		
_____	_____	_____	_____
	_____		
_____	_____	_____	_____
	_____		
_____	_____	_____	_____
	_____		
_____	_____	_____	_____
	_____		
_____	_____	_____	_____
	_____		
_____	_____	_____	_____
	_____		
_____	_____	_____	_____
	_____		
_____	_____	_____	_____
	_____		
_____	_____	_____	_____
	_____		

ADVANCE STUDY ASSIGNMENT: Identification of Unknown Solids

1.   Devise one- or two-step procedures to determine which member of the pair is present in an unknown. For each procedure write the net ionic equation for the identifying reaction.

   a. Na_2CO_3 or $NaCl$

   b. $NiCl_2$ or $BaCl_2$

   c. KCl or NH_4Cl

   d. $NaOH$ or $Mg(OH)_2$

   e. $KC_2H_3O_2$ or $KSCN$

   f. $NaHSO_4$ or Na_2SO_4

   g. $MgCO_3$ or $BaSO_4$

   h. $ZnCO_3$ or $CdCO_3$

   i. $AgCl$ or Ag_2O

   j. ZnS or CuS

EXPERIMENT

37 • Qualitative Analysis of Group I Cations

PRECIPITATION AND SEPARATION OF GROUP I IONS

The chlorides of Pb^{2+}, Hg_2^{2+}, and Ag^+ are all insoluble in cold water. They can be removed as a group from solution by the addition of HCl. The reactions that occur are simple precipitations and can be represented by the equations:

$$Ag^+(aq) + Cl^-(aq) \rightarrow AgCl(s) \tag{1}$$

$$Pb^{2+}(aq) + 2\,Cl^-(aq) \rightarrow PbCl_2(s) \tag{2}$$

$$Hg_2^{2+}(aq) + 2\,Cl^-(aq) \rightarrow Hg_2Cl_2(s) \tag{3}$$

It is important to add enough HCl to ensure complete precipitation, but not too large an excess. In concentrated HCl solution, these chlorides tend to dissolve, producing chloro-complexes such as $AgCl_2^-$.

Lead chloride is separated from the other two chlorides by heating with water. The $PbCl_2$ dissolves in hot water by the reverse of Reaction 2:

$$PbCl_2(s) \rightarrow Pb^{2+}(aq) + 2\,Cl^-(aq) \tag{4}$$

Once Pb^{2+} has been put into solution, we can check for its presence by adding a solution of K_2CrO_4. The chromate ion, CrO_4^{2-}, gives a yellow precipitate with Pb^{2+}:

$$Pb^{2+}(aq) + CrO_4^{2-}(aq) \rightarrow PbCrO_4(s) \tag{5}$$
$$\text{yellow}$$

The other two insoluble chlorides, AgCl and Hg_2Cl_2, can be separated by adding aqueous ammonia. Silver chloride dissolves, forming the complex ion $Ag(NH_3)_2^+$:

$$AgCl(s) + 2\,NH_3(aq) \rightarrow Ag(NH_3)_2^+(aq) + Cl^-(aq) \tag{6}$$

Ammonia also reacts with Hg_2Cl_2 via a rather unusual oxidation-reduction reaction. The products include finely divided metallic mercury, which is black, and a compound of formula $HgNH_2Cl$, which is white:

$$Hg_2Cl_2(s) + 2\,NH_3(aq) \rightarrow Hg(l) + HgNH_2Cl(s) + NH_4^+(aq) + Cl^-(aq) \tag{7}$$
$$\text{white} \qquad\qquad\qquad \text{black} \quad\; \text{white}$$

As this reaction occurs, the solid appears to change color, from white to black or grey.

The solution containing $Ag(NH_3)_2^+$ needs to be further tested to establish the presence of silver. The addition of a strong acid (HNO_3) to the solution destroys the complex ion and reprecipitates silver chloride. We may consider that this reaction occurs in two steps:

$$Ag(NH_3)_2^+(aq) + 2\,H^+(aq) \rightarrow Ag^+(aq) + 2\,NH_4^+(aq)$$
$$\underline{Ag^+(aq) + Cl^-(aq) \rightarrow AgCl(s)}$$

$$Ag(NH_3)_2^+(aq) + 2\,H^+(aq) + Cl^-(aq) \rightarrow AgCl(s) + 2\,NH_4^+(aq) \tag{8}$$
$$\text{white}$$

EXPERIMENTAL PROCEDURE

1. Precipitation of Group I Ions. To gain familiarity with the analysis scheme we will first analyze a known Group I solution, made by mixing equal volumes of 0.1 M $AgNO_3$, 0.2 M $Pb(NO_3)_2$, and 0.1 M $Hg_2(NO_3)_2$.

Add 2 drops of 6 M HCl to 1 ml of the known solution in a small test tube. Centrifuge the solution, being careful to place a blank test tube containing an equal volume of water in the opposite tube of the centrifuge. Add one more drop of the 6 M HCl to the solution to test for completeness of precipitation. Centrifuge again if necessary, and decant the supernatant solution from the chloride precipitate. The solution should be saved for further study if ions from other groups may be present.

2. Separation of Pb^{2+}. Wash the precipitate with 1 or 2 ml of water. Stir with a glass rod, centrifuge, and decant. Discard the decanted liquid.

Add 2 ml of distilled water to the precipitate in the test tube and place in a 250-ml beaker which is half full of boiling water. Allow the tube to remain in the boiling water bath for a few minutes, and stir occasionally with a glass rod.

Centrifuge the hot solution and quickly pour it into another test tube. Save the remaining precipitate for further study.

3. Identification of Pb^{2+}. Add one drop of 6 M acetic acid and a few drops of 1 M K_2CrO_4 to the solution from 2. If Pb^{2+} is present in the solution, a yellow precipitate of $PbCrO_4$ will form.

4. Separation and Identification of Hg_2^{2+}. Add 10 drops of 6 M NH_3 to the precipitate from 2 and stir thoroughly. Centrifuge the solution and decant. A gray or black precipitate, produced by reaction of Hg_2Cl_2 with ammonia to produce metallic mercury, will establish the presence of Hg_2^{2+}.

5. Identification of Ag^+. Add 6 M HNO_3 to the solution from 4 until it is acidic toward litmus paper. Test for acidity by dipping the end of your stirring rod in the solution and then touching it to a piece of blue litmus paper (red in acid solution). If Ag^+ is present in the acidified solution, a white precipitate of AgCl will form.

6. When you have completed the tests on the known solution, obtain an unknown and analyze it for the possible presence of Ag^+, Pb^{2+}, and Hg_2^{2+}.

FLOW DIAGRAMS

It is possible to summarize the directions for analysis of the Group I cations in what is called a flow diagram. In the diagram, successive steps in the procedure are linked with arrows. Reactant cations or reactant substances containing the ions are at one end of each arrow and products formed are at the other end. Reagents and conditions used to carry out each step are placed alongside the arrows. A partially completed flow diagram for the Group I ions follows:

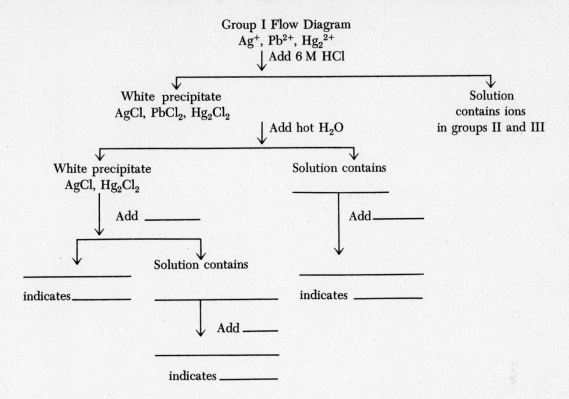

Group I Flow Diagram
Ag^+, Pb^{2+}, Hg_2^{2+}

You will find it useful to construct flow diagrams for each of the cation groups. You can use such diagrams in the laboratory to serve as a brief guide to procedure, and you can use them to directly record your observations on your known and unknown solutions.

OBSERVATIONS AND REPORT SHEET: **Qualitative Analysis of**
                                          Group I Cations

Flow Diagram for Group I

Observations on known (record on diagram if different from those on prepared diagram).

Observations on unknown (record on diagram in colored pencil to distinguish from observations on known).

Unknown no. _____

Ions reported present _____ _____ _____

ADVANCE STUDY ASSIGNMENT: Group I Cations

1. On the report sheet, complete the flow diagram for the separation and identification of the ions in Group I.

2. Write balanced net ionic equations for the following reactions:
   a. The precipitation of the chloride of Pb^{2+} in Step 1.

   b. The formation of a yellow precipitate in Step 3.

   c. The formation of a black precipitate in Step 4.

   d. The reaction that AgCl undergoes in Step 4.

3. A solution may contain Ag^+, Pb^{2+}, and Hg_2^{2+}. A white precipitate forms on addition of 6 M HCl. The precipitate is insoluble in hot water and turns black on addition of 6 M NH_3. Which of the ions are present, which are absent, and which remain undetermined? State your reasoning.

Present _____

Absent _____

In doubt _____

EXPERIMENT

38 • Qualitative Analysis of Group II Cations

PRECIPITATION AND SEPARATION OF GROUP II IONS

The sulfides of the four Group II ions, Bi^{3+}, Sn^{4+}, Sb^{3+}, and Cu^{2+}, are insoluble at a pH of 0.5. The solution is adjusted to this pH and then saturated with H_2S, which precipitates Bi_2S_3, SnS_2, Sb_2S_3 and CuS. The reaction with Bi^{3+} is typical:

$$2\,Bi^{3+}(aq) + 3\,H_2S(aq) \rightarrow Bi_2S_3(s) + 6\,H^+(aq) \tag{1}$$
$$\text{black}$$

Saturation with H_2S could be achieved by simply bubbling the gas from a generator through the solution. A more convenient method, however, is to heat the acid solution after adding a small amount of thioacetamide. This compound, CH_3CSNH_2, hydrolyzes when heated in water solution to liberate H_2S:

$$CH_3CSNH_2(aq) + 2\,H_2O \rightarrow H_2S(aq) + CH_3COO^-(aq) + NH_4^+(aq) \tag{2}$$

Using thioacetamide as the precipitating reagent has the advantage of minimizing odor problems and giving denser precipitates.

The four insoluble sulfides can be separated into two subgroups by extracting with a solution of sodium hydroxide. The sulfides of tin and antimony dissolve, forming hydroxo-complexes:

$$SnS_2(s) + 6\,OH^-(aq) \rightarrow Sn(OH)_6^{2-}(aq) + 2\,S^{2-}(aq) \tag{3}$$

$$Sb_2S_3(s) + 8\,OH^-(aq) \rightarrow 2\,Sb(OH)_4^-(aq) + 3\,S^{2-}(aq) \tag{4}$$

Since Cu^{2+} and Bi^{3+} do not readily form hydroxo-complexes, CuS and Bi_2S_3 do not dissolve in solutions of NaOH.

The solution containing the $Sb(OH)_4^-$ and $Sn(OH)_6^{2-}$ complex ions is treated with HCl and thioacetamide. The H^+ ions of the strong acid HCl destroy the hydroxo-complexes; the free cations then reprecipitate as the sulfides. The reaction with $Sn(OH)_6^{2-}$ may be written as:

$$Sn(OH)_6^{2-}(aq) + 6\,H^+(aq) \rightarrow Sn^{4+}(aq) + 6\,H_2O$$

$$Sn^{4+}(aq) + 2\,H_2S(aq) \rightarrow SnS_2(s) + 4\,H^+(aq)$$

$$\overline{Sn(OH)_6^{2-}(aq) + 2\,H^+(aq) + 2\,H_2S(aq) \rightarrow SnS_2(s) + 6\,H_2O} \tag{5}$$
$$\text{tan}$$

The $Sb(OH)_4^-$ ion behaves in a very similar manner, being converted first to Sb^{3+} and then to Sb_2S_3. The Sb_2S_3 and SnS_2 are then dissolved as chloro-complexes in hydrochloric acid and their presence is confirmed by appropriate tests.

The confirmatory test for tin takes advantage of the two oxidation states, +2 and +4, of the metal. Aluminum is added to reduce Sn^{4+} to Sn^{2+}:

$$2\,Al(s) + 3\,Sn^{4+}(aq) \rightarrow 2\,Al^{3+}(aq) + 3\,Sn^{2+}(aq) \tag{6}$$

The presence of Sn^{2+} in this solution is detected by adding mercury(II) chloride, $HgCl_2$. This brings about another oxidation-reduction reaction:

$$Sn^{2+}(aq) + 2\,Hg^{2+}(aq) + 2\,Cl^-(aq) \rightarrow Sn^{4+}(aq) + Hg_2Cl_2(s) \qquad (7)$$
$$\text{white}$$

Formation of a white precipitate of insoluble Hg_2Cl_2 confirms the presence of tin. (The precipitate may be greyish because of the formation of finely divided Hg by an oxidation-reduction reaction similar to Reaction 7.)

The Sb^{3+} ion is difficult to confirm in the presence of Sn^{4+}; the colors of the sulfides of these two ions are similar. To prevent interference by Sn^{4+}, the solution to be tested for Sb^{3+} is first treated with oxalic acid. This forms a very stable oxalato-complex with Sn^{4+}, $Sn(C_2O_4)_3{}^{2-}$. Treatment with H_2S then gives a bright orange precipitate of Sb_2S_3 if antimony is present:

$$2\,Sb^{3+}(aq) + 3\,H_2S(aq) \rightarrow Sb_2S_3(s) + 6\,H^+(aq) \qquad (8)$$
$$\text{orange}$$

As pointed out earlier, CuS and Bi_2S_3 are insoluble in NaOH solutions, and they don't dissolve in hydrochloric acid. However, these two sulfides can be brought into solution by treatment with the oxidizing acid, HNO_3. The reactions that occur are of the oxidation-reduction type. The $NO_3{}^-$ ion is reduced, usually to NO_2; S^{2-} ions are oxidized to elementary sulfur and the cation, Cu^{2+} or Bi^{3+}, is brought into solution. The equations are:

$$CuS(s) + 4\,H^+(aq) + 2\,NO_3{}^-(aq) \rightarrow Cu^{2+}(aq) + S(s) + 2\,NO_2(g) + 2\,H_2O \qquad (9)$$

$$Bi_2S_3(s) + 12\,H^+(aq) + 6\,NO_3{}^-(aq) \rightarrow 2\,Bi^{3+}(aq) + 3S(s) + 6\,NO_2(g) + 6\,H_2O \qquad (10)$$

The two ions, Cu^{2+} and Bi^{3+}, are easily separated by the addition of aqueous ammonia. The Cu^{2+} ion is converted to the deep-blue complex, $Cu(NH_3)_4{}^{2+}$:

$$Cu^{2+}(aq) + 4\,NH_3(aq) \rightarrow Cu(NH_3)_4{}^{2+}(aq) \qquad (11)$$
$$\text{deep blue}$$

The reaction of ammonia with Bi^{3+} is quite different. The OH^- ions produced by the reaction of NH_3 with water precipitate Bi^{3+} as $Bi(OH)_3$. We can consider that the reaction occurs in two steps:

$$3\,NH_3(aq) + 3\,H_2O \rightleftarrows 3\,NH_4{}^+(aq) + 3\,OH^-(aq)$$

$$Bi^{3+}(aq) + 3\,OH^-(aq) \rightarrow Bi(OH)_3(s)$$

$$Bi^{3+}(aq) + 3\,NH_3(aq) + 3\,H_2O \rightarrow Bi(OH)_3(s) + 3\,NH_4{}^+(aq) \qquad (12)$$
$$\text{white}$$

To confirm the presence of Bi^{3+}, the precipitate of $Bi(OH)_3$ is dissolved by treating with hydrochloric acid:

$$Bi(OH)_3(s) + 3\,H^+(aq) \rightarrow Bi^{3+}(aq) + 3\,H_2O \qquad (13)$$

The solution formed is poured into distilled water. If Bi^{3+} is present, a white precipitate of bismuth oxychloride, BiOCl, will form:

$$Bi^{3+}(aq) + H_2O + Cl^-(aq) \rightarrow BiOCl(s) + 2\,H^+(aq) \qquad (14)$$
$$\text{white}$$

EXPERIMENTAL PROCEDURE

1. Adjustment of pH Prior to Precipitation. Obtain a 1-ml sample of the "known" solution for Group II, containing equal volumes of 0.1 M solutions of the nitrates or chlorides of Sn^{4+}, Sb^{3+}, Cu^{2+}, and Bi^{3+}. The Group II known will be very acidic due to addition of HCl, which is required to keep the salts of Bi^{3+}, Sn^{4+}, and Sb^{3+} in solution. Add 6 M NH_3 until the solution, after stirring, is just basic to litmus paper (blue in base). A precipitate will probably form during this step because of formation of insoluble oxychlorides of bismuth, tin, or antimony. Then add 1 drop of 6 M HCl for each ml of solution. This should bring the solution to a pH of about 0.5. This can be verified by the use of methyl violet indicator paper with 0.3 M HCl solution (pH = 0.5) as a standard for comparison. This paper can be prepared by placing methyl violet solution on filter paper and allowing it to dry. The 0.3 M HCl solution will give the methyl violet spot a blue-green color. Adjust the pH of your test solution if necessary by adding HCl or NH_3 so that it gives about the same color. Don't forget to stir before testing the pH.

2. Precipitation of Group II Sulfides. Add 15 to 20 drops of 1 M thioacetamide solution to your test solution and heat it with occasional stirring in a boiling water bath for 5 minutes. Centrifuge the solution, which will contain the Group III cations if they are present, and decant into a second test tube. Test the solution for completeness of precipitation by adding 2 drops of thioacetamide and allowing it to stand for 1 minute. If a precipitate forms, add a few more drops of thioacetamide solution and heat again in the boiling water bath. Combine the two batches of precipitate. Wash the precipitate with 2 ml 1 M NH_4Cl solution, stir thoroughly and warm in the water bath; centrifuge, and discard the wash solution.

3. Separation of Group II Sulfides into Two Subgroups. To the sulfide precipitate add 2 ml 1 M NaOH. Heat the test tube in the water bath for 2 minutes, stirring occasionally. Centrifuge and decant the solution into a test tube. Wash any remaining precipitate once with 2 ml water and 1 ml 1 M NaOH and once with 3 ml water. Stir, centrifuge, and discard the wash each time. Add 2 ml 6 M HNO_3 to the precipitate and put the test tube aside (Step 8). At this point, the tin and antimony will be present in the solution as complex ions and the bismuth and copper will be in the sulfide precipitate.

4. Reprecipitation of SnS_2 and Sb_2S_3. Add 6 M HCl to the solution from Step 3 until, after stirring, it is just acidic to litmus. Upon acidification, some of the tin and antimony will again precipitate as orange sulfide. Add 5 drops 1 M thioacetamide and heat in the water bath as before to complete the precipitation. Centrifuge and decant the solution, which may be discarded.

5. Dissolving SnS_2 and Sb_2S_3 in Acid Solution. Add 2 ml 6 M HCl to the precipitate from Step 4, and heat in the water bath for a minute or two to dissolve the precipitate. Transfer the solution to a 30-ml beaker, and boil it, gently, for about a minute, to drive out the H_2S. Add 1 ml water and pour the liquid back into a small test tube; centrifuge out any insoluble sulfide residue.

6. Confirmation of the Presence of Tin. Pour about half of the solution from Step 5 into a test tube and add 2 ml 6 M HCl and a piece of aluminum wire. The aluminum will react with the acid, producing H_2 gas and reducing any Sn^{4+} present to Sn^{2+}. Heat the mixture in a water bath. A black precipitate is highly indicative of the presence of antimony. Heat the tube for 2 minutes after all of the wire has reacted. Centrifuge, and decant the solution into a test tube; discard any solid. Add 2 or 3 drops of 0.1 M $HgCl_2$ to the solution. If tin is present, an oxidation-reduction reaction will occur, which slowly produces a white or gray precipitate of Hg_2Cl_2 or Hg. This precipitate confirms the presence of tin.

7. Confirmation of the Presence of Antimony. To the other half of the solution from Step 5, add 5 ml water and about 0.5 g of oxalic acid. Heat the mixture, if necessary, in the water bath to dissolve the acid. This reagent forms a very stable complex with the Sn^{4+} ion. Add 10 drops 1 M thioacetamide and heat in the boiling water bath. The formation of an orange precipitate of Sb_2S_3 confirms the presence of antimony.

8. Dissolving the CuS and Bi_2S_2. Heat the test tube containing the sulfide precipitate from Step 3 in the water bath. Any sulfides that have not already dissolved should go into solution within a minute or two. Centrifuge to remove any insoluble sulfur, and decant the liquid into a 30-ml beaker. Boil off, carefully, about half of the liquid.

9. Confirmation of the Presence of Copper. Pour the liquid from Step 8 into a small test tube and add 6 M NH_3 dropwise until the mixture is just basic to litmus. Add 5 drops more. Centrifuge out any white precipitate which forms. If the solution is deep blue, the color is due to $Cu(NH_3)_4{}^{2+}$ ion and establishes the presence of copper.

10. Confirmation of the Presence of Bismuth. Decant the solution from Step 9, and to the solid, which probably contains bismuth, add 5 drops 6 M HCl. The solid should dissolve readily, as a bismuth chloride complex ion. Pour the solution into a 600-ml beaker two-thirds full of distilled water. A white cloudiness, caused by the slow precipitation of BiOCl, confirms the presence of bismuth.

11. When you have completed the analysis of your known, obtain a Group II unknown and test it for the presence of Sn^{4+}, Sb^{3+}, Cu^{2+}, and Bi^{3+}.

OBSERVATIONS AND REPORT SHEET: Qualitative Analysis of
Group II Cations

Flow Diagram for Group II

Observations on known (record on diagram if different from those on prepared diagram).

Observations on unknown (record on diagram in colored pencil to distinguish from observations on known).

Unknown no. _____

Ions reported present _____   _____   _____   _____   **305**

ADVANCE STUDY ASSIGNMENT: Qualitative Analysis of
Group II Cations

1. Prepare a complete flow diagram for the separation and identification of the Group II cations and put it on your report sheet.

2. Write balanced net ionic equations for the following reactions:

   a. Precipitation of the tin(IV) sulfide with H_2S.

   b. The confirmatory test for antimony.

   c. The confirmatory test for bismuth.

   d. The dissolving of Bi_2S_3 in hot nitric acid.

3. A solution which may contain Cu^{2+}, Bi^{3+}, Sn^{4+}, or Sb^{3+} ions is treated with thioacetamide in an acid medium. The black precipitate that forms is partly soluble in strongly alkaline solution. The precipitate that remains is soluble in 6 M HNO_3 and gives only a blue solution on treatment with excess NH_3. The alkaline solution, when acidified, produces an orange precipitate. On the basis of this information, which ions are present, which are absent, and which are still in doubt?

Present _____

Absent _____

In doubt _____

EXPERIMENT

39 • Qualitative Analysis of Group III Cations

PRECIPITATION AND SEPARATION OF THE GROUP III IONS

Four ions in Group III are Cr^{3+}, Al^{3+}, Fe^{3+}, and Ni^{2+}. The first step in the analysis involves treating the solution with sodium hydroxide, NaOH, and sodium hypochlorite, NaOCl. The OCl^- ion oxidizes Cr^{3+} to CrO_4^{2-}:

$$2\,Cr^{3+}(aq) + 3\,OCl^-(aq) + 10\,OH^-(aq) \rightarrow 2\,CrO_4^{2-}(aq) + 3\,Cl^-(aq) + 5\,H_2O \qquad (1)$$
$$\text{yellow}$$

The chromate ion, CrO_4^{2-}, stays in solution. The same is true of the hydroxo-complex ion $Al(OH)_4^-$, formed by the reaction of Al^{3+} with excess OH^-:

$$Al^{3+}(aq) + 4\,OH^-(aq) \rightarrow Al(OH)_4^-(aq) \qquad (2)$$

In contrast, the other two ions in the group form insoluble hydroxides under these conditions:

$$Ni^{2+}(aq) + 2\,OH^-(aq) \rightarrow Ni(OH)_2(s) \qquad (3)$$
$$\text{green}$$

$$Fe^{3+}(aq) + 3\,OH^-(aq) \rightarrow Fe(OH)_3(s) \qquad (4)$$
$$\text{red}$$

The Ni^{2+} and Fe^{3+} ions, unlike Al^{3+}, do not readily form hydroxo-complexes. Unlike Cr^{3+}, they do not have a stable higher oxidation state and so are not oxidized by ClO^-.

To separate aluminum from chromium, the solution containing CrO_4^{2-} and $Al(OH)_4^-$ is first acidified. This destroys the hydroxo-complex of aluminum:

$$Al(OH)_4^-(aq) + 4\,H^+(aq) \rightarrow Al^{3+}(aq) + 4\,H_2O \qquad (5)$$

Treatment with aqueous ammonia then gives a white gelatinous precipitate of aluminum hydroxide. We can think of this reaction as occurring in two steps:

$$3\,NH_3(aq) + 3\,H_2O \rightleftharpoons NH_4^+(aq) + 3\,OH^-(aq)$$

$$Al^{3+}(aq) + 3\,OH^-(aq) \rightarrow Al(OH)_3(s)$$

$$\overline{}$$

$$Al^{3+}(aq) + 3\,NH_3(aq) + 3\,H_2O \rightarrow Al(OH)_3(s) + 3\,NH_4^+(aq) \qquad (6)$$
$$\text{white}$$

The concentration of OH^- in dilute NH_3 is too low to form the $Al(OH)_4^-$ complex ion by Reaction 2.

The CrO_4^{2-} ion remains in solution after Al^{3+} has been precipitated. It can be tested for by precipitation as yellow $BaCrO_4$ by the addition of $BaCl_2$ solution:

$$Ba^{2+}(aq) + CrO_4^{2-}(aq) \rightarrow BaCrO_4(aq) \qquad (7)$$
$$\text{lt. yellow}$$

The precipitate of $BaCrO_4$ is dissolved in acid; the solution formed is then treated with hydrogen peroxide, H_2O_2. A deep-blue color is produced, due to the presence of a peroxo-compound, probably CrO_5. The reaction may be represented by the overall equation:

$$2\,BaCrO_4(s) + 4\,H^+(aq) + 4\,H_2O_2(aq) \rightarrow 2\,Ba^{2+}(aq) + 2\,CrO_5(aq) + 6\,H_2O \qquad (8)$$
$$\text{blue}$$

The mixed precipitate of $Ni(OH)_2$ and $Fe(OH)_3$ formed by Reactions 3 and 4 is dissolved by adding a strong acid, HNO_3. An acid-base reaction occurs:

$$Ni(OH)_2(s) + 2\,H^+(aq) \rightarrow Ni^{2+}(aq) + 2\,H_2O \qquad (9)$$

$$Fe(OH)_3(s) + 3\,H^+(aq) \rightarrow Fe^{3+}(aq) + 3\,H_2O \qquad (10)$$

At this point, the Ni^{2+} and Fe^{3+} ions are separated by adding ammonia. The Ni^{2+} ion is converted to the deep-blue complex $Ni(NH_3)_6^{2+}$, which stays in solution:

$$Ni^{2+}(aq) + 6\,NH_3(aq) \rightarrow Ni(NH_3)_6^{2+}(aq) \qquad (11)$$
$$\text{blue}$$

while the Fe^{3+} ion, which does not readily form a complex with NH_3, is reprecipitated as $Fe(OH)_3$:

$$3\,NH_3(aq) + 3\,H_2O \rightleftharpoons 3\,NH_4^+(aq) + 3\,OH^-(aq)$$

$$Fe^{3+}(aq) + 3\,OH^-(aq) \rightarrow Fe(OH)_3(s)$$

$$Fe^{3+}(aq) + 3\,NH_3(aq) + 3\,H_2O \rightarrow Fe(OH)_3(s) + 3\,NH_4^+(aq) \qquad (12)$$
$$\text{red}$$

The confirmatory test for nickel in the solution is made by adding an organic reagent, dimethylglyoxime, $C_4H_8N_2O_2$. This gives a deep rose-colored precipitate with nickel:

$$Ni^{2+}(aq) + 2\,C_4H_8N_2O_2(aq) \rightarrow Ni(C_4H_7N_2O_2)_2(s) + 2\,H^+(aq) \qquad (13)$$
$$\text{rose}$$

We can confirm Fe^{3+} by dissolving the precipitate of $Fe(OH)_3$ in HCl (Reaction 10) and adding KSCN solution. If iron(III) is present, the blood-red complex $FeSCN^{2+}$ will form:

$$Fe^{3+}(aq) + SCN^-(aq) \rightarrow FeSCN^{2+}(aq) \qquad (14)$$
$$\text{red}$$

EXPERIMENTAL PROCEDURE

1. If you are testing a solution from which Group II cations have been precipitated (Experiment 38), remove the H_2S and excess acid by boiling the solution in a small beaker until the volume is reduced to about 1 ml. Remove any sulfur residue by centrifuging the solution.

If you are working on the analysis of Group III ions only, prepare a known solution containing Fe^{3+}, Al^{3+}, Cr^{3+}, and Ni^{2+} by mixing together 5-drop portions of each of the appropriate 0.1 M solutions.

2. Oxidation of Cr(III) to Cr(VI) and Separation of Insoluble Hydroxides. Add 1 ml of 6 M NaOH to 1 ml of the known solution in a 30-ml beaker. Boil very gently for a minute, stirring to minimize bumping. Remove heat, and slowly add 1 ml of 1 M NaClO, sodium hypochlorite. Swirl the beaker for 30 seconds, using your tongs if necessary. Then boil the mixture gently for one minute. Add 8 drops 6 M NH_3 and boil gently for another minute. Transfer the mixture to a test tube and centrifuge out the solid, which contains iron and nickel hydroxides. Decant the solution, which contains chromium and aluminum (CrO_4^{2-} and $Al(OH)_4^-$) ions into a test tube. Wash the solid with 2 ml water and 10 drops 6 M NaOH; stir well, centrifuge, and discard the wash. Add 1 ml 6 M HNO_3 to the solid and put the tube aside (Step 6).

3. Separation of Al from Cr.   Acidify the solution from Step 2 by adding 6 M acetic acid slowly until, after stirring, the mixture is definitely acidic to litmus. Add 6 M NH_3, drop by drop, until the solution is basic to litmus; then add 10 drops in excess. Stir the mixture for a minute or so to bring the system to equilibrium. If aluminum is present, a light translucent, gelatinous white precipitate of $Al(OH)_3$ should be floating in the clear (possibly yellow) solution. Centrifuge out the solid and transfer the liquid to a test tube. Wash the solid with 3 ml water, while warming the test tube in the water bath and stirring well. Centrifuge and discard the wash.

4. Confirmation of the Presence of Aluminum.   To the precipitate from Step 3 add 2 drops 6 M acetic acid and 3 ml water. Stir to dissolve the solid, and to the resulting solution add 2 drops catechol violet reagent. Stir. If the solution obtained is blue, aluminum is present.

5. Confirmation of the Presence of Chromium.   If the solution from Step 3 is yellow, chromium is probably present. To that solution, add 6 drops 1 M $BaCl_2$ and stir. In the presence of chromium you obtain a very finely divided yellow precipitate of $BaCrO_4$. Put the test tube in the boiling water bath for a few minutes; then centrifuge out the solid, discarding the liquid. Wash the solid with 2 ml water; centrifuge and discard the wash. To the solid add 5 drops 6 M HNO_3, and stir to dissolve the $BaCrO_4$. Add 1 ml water, stir the orange solution and add 2 drops of 3 per cent H_2O_2. A deep blue solution, whose color may fade quite rapidly, is confirmatory evidence for the presence of chromium.

6. Separation of Ni^{2+} and Fe^{3+}.   Returning to the precipitate obtained in Step 2, stir to dissolve the solid in the HNO_3. If necessary, warm the test tube in the water bath to complete the solution process. Then add 6 M NH_3 until the solution is basic to litmus. At that point iron will reprecipitate as brown $Fe(OH)_3$. Add 10 more drops of the 6 M NH_3 and stir to bring all of the nickel into solution as $Ni(NH_3)_6^{2+}$ ion. Centrifuge and decant the solution into a test tube.

7. Confirmation of the Presence of Nickel.   Add 4 drops of dimethylglyoxime reagent to the solution from Step 6. A rose-red precipitate establishes the presence of nickel.

8. Confirmation of the Presence of Iron.   Dissolve the precipitate from Step 6 in 6 drops 6 M HCl. Add 2 ml water and stir. Then add 2 drops 0.5 M KSCN. Formation of a deep red solution of $FeSCN^{2+}$ is a definitive test for iron.

9.   When you have completed your analysis of the known solution, obtain a Group III unknown and test it for the possible presence of Al^{3+}, Cr^{3+}, Ni^{2+}, and Fe^{3+}.

Name _____ Section _____

OBSERVATIONS AND REPORT SHEET: Qualitative Analysis of
Group III Cations

Flow Diagram for Group III

Observation on known (record on diagram if different from those on prepared diagram)

Observations on unknown (record on diagram in colored pencil to distinguish from observations on known)

Unknown no. _____

Ions reported present _____   _____   _____   _____

Name _____ **Section** _____

ADVANCE STUDY ASSIGNMENT: **Qualitative Analysis of**
                                 Group III Cations

1.   Prepare a flow diagram for the separation and identification of the ions in Group III and put it on your report sheet.

2.   Write balanced net ionic equations for the following reactions:

a. Dissolving of $Fe(OH)_3$ in nitric acid.

b. Oxidation of Cr^{3+} to CrO_4^{2-} by ClO^- in alkaline solution (ClO^- is converted to Cl^-).

c. The confirmatory test for Ni^{2+}.

d. The confirmatory test for Fe^{3+}.

3.   A solution may contain any of the Group III cations. Treatment of the solution with ClO^- in alkaline medium yields a yellow solution and a colored precipitate. The acidified solution is unaffected by treatment with NH_3. The colored precipitate dissolves in nitric acid; addition of excess NH_3 to this acid solution produces only a blue solution. On the basis of this information, which Group III cations are present, absent, or still in doubt?

Present _____

Absent _____

In doubt _____

EXPERIMENT

40 • The Ten Test Tube Mystery

Having carried out at least some of the previous experiments on qualitative analysis, you should have acquired some familiarity with the properties of the cations and anions you studied. Along with this you should at this point have had some experience with interpretation of the data in Appendix II on the solubilities of ionic substances in different media.

In this experiment we are going to ask you to apply what you have learned to a related but somewhat different kind of problem. You will be furnished ten numbered test tubes in each of which there will be a solution of a single substance. You will be provided, one week in advance, with a list giving the formula and molarity of each of the ten solutes that will be used. Your problem in the lab will be to find out which solution is in which test tube—that is, to assign a test tube number to each of the solution compositions. You are to do this by intermixing small volumes of the solutions in the test tubes. No external reagents or acid-base indicators such as litmus are allowed. You are permitted, however, to use the odor and color of the different solute species and to make use of heat effects in reactions in your system for identification. Each student will have a different set of test tube number-solution composition correlations, and there will be several different sets of compositions as well.

Of the ten solutions, four are common laboratory reagents. They are 6 M HCl, 3 M H_2SO_4, 6 M NH_3, and 6 M NaOH. The other solutions will usually be 0.1 M nitrate, chloride, or sulfate solutions of the cations which have been studied in previous experiments in this manual (Experiments 11, 35, 37, 38, and 39).

In order to determine which solution is in each test tube, you will need to know what happens when the various solutions are mixed, one with another. In some cases, nothing happens that you can observe. This will often be the case when a solution containing one of the cations is mixed with a solution of another. When one of the reagents is mixed with a cation solution you may get a precipitate, white or colored, and that precipitate may dissolve in excess reagent by complex ion formation. In a few cases a gas may be evolved. When one laboratory reagent is mixed with another, you may find that the resulting solution gets very hot and/or that a visible vapor is produced.

There is no way that you will be able to solve your particular test tube mystery without doing some preliminary work. You will need to know what to expect when any two of your ten solutions are mixed. You can find this out, for any pair, by scrutiny of Appendix II, by consulting your chemistry text, and by referring to various reference works on qualitative analysis.° A convenient way to tabulate the information you obtain is to set up a matrix with ten columns and ten rows, one for each solution. The key information about a mixture of two solutions is put in the space where the row for one solution and the column for the other intersect (as is done in Appendix II for various cations and anions). In that space you might put an NR, to indicate no apparent reaction on mixing of the two solutions. If a precipitate forms, put a P, followed by a D if the precipitate dissolves in excess reagent. If the precipitate or final solution is colored, state the color. If heat is evolved, write an H; if a gas or smoke is formed, put a G or S. Since mixing solution A with B is the same as mixing B with A, not all 100 spaces in the 10-by-10 matrix need to be filled. Actually, there are only 45 possible different pairs, since A with A is not very informative either.

Since you are allowed to use the odor or color of a solution to identify it, the problem is somewhat simpler than it might first appear. In each set of ten solutions you will probably be able

---

° For example, the following references may be useful:

1. *A Textbook of Macro and Semimicro Qualitative Inorganic Analysis,* 4th ed., by A. I. Vogel. Longmans, Green, and Co., 1954.

2. *Chemical Principles, Alternate Edition,* by W. L. Masterton, E. J. Slowinski, and C. L. Stanitski. Philadelphia, Saunders College Publishing, 1983.

3. *Handbook of Chemistry and Physics.* Akron, Chemical Rubber Publishing Co., 1984.

to identify at least two solutions by odor and color tests. Knowing those solutions, you can make mixtures with the other solutions in which one of the components is known. From the results obtained with those mixtures, and the information in the matrix, you can identify other solutions. These can be used to identify still others, until finally the entire set of ten is identified unequivocally.

Let us now go through the various steps that would be involved in the solution to a somewhat simpler problem than the one you will solve. There are several steps, including constructing the reaction matrix, identifying solutions by simple observations, and a rationale for efficient mixing tests. Let us assume we have to identify the following six solutions in numbered test tubes:

0.1 M $NiSO_4$	0.1 M $BiCl_3$ (in 3 M HCl)
0.1 M $BaCl_2$	6 M NaOH
0.1 M $Al(NO_3)_3$	3 M H_2SO_4

Construction of Reaction Matrix.   In the matrix there will be six rows and six columns, one for each of the solutions. The matrix would be set up as in Table 40.1.

TABLE 40.1   REACTION MATRIX FOR SIX SOLUTIONS

	$Al(NO_3)_3$	$BaCl_2$	$NiSO_4$	$BiCl_3$	NaOH	H_2SO_4
$Al(NO_3)_3$		NR	NR	NR	P, D	NR
$BaCl_2$			P	NR	sl P	P
$NiSO_4$(green)				NR	P (green)	NR
$BiCl_3$					P, H	NR
NaOH						H
H_2SO_4						

Each solution contains a cation and an anion, both of which need to be considered. We need to include every possible pair of solutions; for six solutions there are 15 possible pairs. So we need 15 pieces of information as to what happens when a pair of solutions is mixed. The first pair we might consider is $Al(NO_3)_3$ and $BaCl_2$, containing Al^{3+}, NO_3^-, Ba^{2+}, and Cl^- ions. Consulting Appendix II, we see that in the Al^{3+}, Cl^- space there is an S, meaning that $AlCl_3$ is soluble and will not precipitate when aluminum and chloride ions are mixed. Similarly, $Ba(NO_3)_3$ is soluble since all nitrates are soluble. This means that there will be no reaction when the solutions of $Al(NO_3)_3$ and $BaCl_2$ are mixed. So, in the space for $Al(NO_3)_3$-$BaCl_2$, we have written NR. The same is true if we mix solutions of $Al(NO_3)_3$ and $NiSO_4$, so we insert NR in that space. Since $BiCl_3$ won't react with $Al(NO_3)_3$ solutions, there is an NR in that space too. However, when $Al(NO_3)_3$ is mixed with NaOH, one of the possible products is $Al(OH)_3$, which we see in Appendix II is insoluble in water, but dissolves in acid and excess OH^- ion (A, B). This means that on adding NaOH to $Al(NO_3)_3$, we would initially get a precipitate of $Al(OH)_3$, P, but that it would dissolve in excess NaOH, D. The precipitate is white and the solution is colorless, so we have just a P and a D in the space for that mixture.

Proceeding now to the $BaCl_2$ row, we don't need to consider the first two columns, since they would give no new information. If $BaCl_2$ and $NiSO_4$ were mixed, $BaSO_4$ is a possible product. In Appendix II we see that $BaSO_4$ is insoluble in all common solvents (I). Hence, on mixing those solutions, we would get a precipitate of $BaSO_4$, P. Using the *Handbook of Chemistry and Physics*, or some other source, we find that it is white, so a P goes into that space. With $BiCl_3$ there would be no reaction; with NaOH we might get a slight precipitate (S^- for $Ba(OH)_2$ in Appendix II), so we put sl P in the appropriate space. With H_2SO_4, we would again expect to get a white precipitate of $BaSO_4$, P.

In the row for $NiSO_4$ we note that the color of the solution is green, since in Appendix II we see that Ni^{2+} (aq) ion is green. The first column is that for $BiCl_3$; no precipitate forms on mixing solutions of $NiSO_4$ and $BiCl_3$, so NR is in that space. With NaOH, $Ni(OH)_2$ would precipitate, since the entry in Appendix II for Ni^{2+}, OH^- is not S. From the *Handbook* or other sources we find that that precipitate is green, so in the space we put a P (green). In Appendix II the $Ni(OH)_2$ entry is A, N, which tells us that the precipitate will dissolve in 6 M NH_3 (but not in 6 M NaOH since there is no B in the space). There would be no reaction between solutions of $NiSO_4$ and H_2SO_4(NR).

With $BiCl_3$ in 3 M HCl we have both a salt and an acid in the solution. Addition of 6 M NaOH should produce a precipitate (white) of $Bi(OH)_3$, P. There would also be a substantial exothermic acid-base reaction between H^+ ions in the HCl and OH^- ions in the NaOH, so there would be a very noticeable rise in temperature on mixing, as denoted by H in the space. There would be no reaction with H_2SO_4.

When 6 M NaOH is mixed with 3 M H_2SO_4 there should be a large heat effect due to the acid-base reaction that occurs, so in the space for NaOH-H_2SO_4 we have an H.

The matrix in Table 40.1 summarizes the reaction information that will be needed for identification of the solutions in the test tubes. Your matrix can be constructed by the reasoning used in making this one.

Identifying Solutions by Simple Observations. When we made the matrix for the six test tube system, we noted that the $NiSO_4$ solution would be green, since Ni^{2+} ion is green. So, as soon as we see the six test tubes, we can pick out the 0.1 M $NiSO_4$. Let us assume that it is in test tube no. 4. We have now identified one solution out of six.

Selection of Efficient Mixing Tests. Knowing that test tube no. 4 contains $NiSO_4$, we mix the solution in test tube no. 4 with all of the other solutions. For each mixture, we record what we observe. The results might be as follows:

4 + 1	no obvious reaction	4 + 5	no obvious reaction
4 + 2	no obvious reaction	4 + 6	white precipitate
4 + 3	green precipitate forms		

Referring to the matrix, we would expect precipitates for mixtures of $NiSO_4$ and NaOH, and $NiSO_4$ and $BaCl_2$. The former precipitate would be green, the latter white. Clearly, the solution in test tube no. 3 is 6 M NaOH. The solution in test tube 6 must be 0.1 M $BaCl_2$. At this point three solutions have been identified.

Since 6 M NaOH reacts with several of the solutions, and since we know now that it is in test tube no. 3, we mix the solution in test tube no. 3 with those that remain unidentified, with the following results:

3 + 1	solution becomes hot
3 + 2	white precipitate, solution gets hot
3 + 5	apparently no reaction

Again, referring to the matrix we would expect heat evolution with mixtures of NaOH and H_2SO_4, and NaOH and the HCl in the $BiCl_3$. In addition, we would get a white precipitate with $BiCl_3$. It is obvious that solution 1 is 3 M H_2SO_4. Solution 2 must be 0.1 M $BiCl_3$ in 3 M HCl. By elimination, solution 5 must be $Al(NO_3)_3$. We repeat the 3 + 5 mixture, since there should have been an initial precipitate which dissolved in excess NaOH. This time, using one drop of the solution 3 added to 1 ml of solution 5, we get a precipitate. In excess of solution 3 the precipitate dissolves as the aluminum hydroxide complex ion is formed.

The approach we have used in our example should be effective for identification of the solutions in your set. After you have tentatively assigned each solution to a test tube, carry out some confirmatory mixing reactions, so that for each solution you have at least two mixtures which give results which support your conclusions.

Wear your safety glasses while performing the experiment.

DATA AND OBSERVATIONS: Ten Test Tube Mystery

Solutions that were Identified Directly

Solution	Observations	Conclusions	Test tube no.

Mixing tests: First Series

Test tube nos.	Observations	Conclusions

Mixing Tests: Second Series

Test tube nos.	Observations	Conclusions

Confirming tests (total of two per solution)

Test tube nos.	Observations	Conclusions

Final identifications: No. 1 _____   No. 2 _____   No. 3 _____

No. 4 _____   No. 5 _____   No. 6 _____   No. 7 _____

No. 8 _____   No. 9 _____   No. 10 _____

ADVANCE STUDY ASSIGNMENT: Ten Test Tube Mystery

1.   Construct the reaction matrix for your set of ten solutions. This should go on the reverse side of this page.

2.   Which solutions should you be able to identify by simple observations?

3.   Outline the procedure you will follow in identifying the solutions that will require mixing tests. Be as specific as you can about what you will look for and what conclusions you will be able to draw from your observations.

EXPERIMENT

41 • Laboratory Examination on Qualitative Analysis of Cations

In Experiments 37 to 39 you worked with a procedure for the qualitative analysis of eleven common cations. By successive application of the procedures used for Groups I, II, and III, you could readily analyze a general unknown for the presence of the eleven ions.

Rather than give you a general unknown, we will give you a more limited unknown, one confined to five possible ions from the group of eleven. Each student will be given a different set of ions to work with and will be assigned his set of ions the week previous to this examination; that week he will be given a set of five letters, corresponding to the five cations which may be in his unknown, according to the following code:

$$A = Ag^+ \quad C = Hg_2^{2+} \quad E = Bi^{3+} \quad G = Sb^{3+} \quad I = Al^{3+} \quad K = Fe^{3+}$$
$$B = Pb^{2+} \quad D = Cu^{2+} \quad F = Sn^{4+} \quad H = Ni^{2+} \quad J = Cr^{3+}$$

Before coming to laboratory for the examination, prepare a flow diagram for the analysis of your five ions. During the examination you will analyze your unknown according to your analysis scheme and will report your results on the page with your flow diagram.

You will be allowed 50 minutes in the laboratory period to complete the analysis. You will receive a bonus if you finish your unknown within 30 minutes and a penalty if it takes you more than 50 minutes. No results will be accepted after 75 minutes.

Your grade on the examination will be based on (1) the accuracy of your analysis, (2) the workability of the scheme for analysis on your flow diagram, (3) the absence of extra steps in your flow sheet, steps that relate to ions other than those that can be in your unknown, and (4) the time it takes you to complete the analysis.

There is to be no communication between students during the examination. During the laboratory period, your laboratory supervisor will not answer any questions concerning the analysis. In preparing your flow diagram you may use the portions of the procedure for Groups I, II, and III, which pertain to the cations that may be in your unknown, or you may use other reactions that you are sure will enable you to make the analysis. It is to your interest to shorten the scheme as much as you can, avoiding any steps that are unnecessary. It is of course important that the scheme you decide upon is one that will work for the set of ions in your unknown.

In developing your analysis scheme there are several precautions you should observe, which are perhaps not at once obvious. Your procedure will in all probability involve removing from the solution first one group of ions, then another, and then perhaps a third. This means that if you are to avoid difficulty in interpretation, the group separation must be *complete:* always test to make sure that the precipitating reagent has been added in sufficient amount to bring down all of the material that should precipitate at that point. If all of the Group I ions are not removed by the addition of an adequate amount of HCl, you may be sure that the remaining Group I ions will precipitate along with Group II when that group is removed from the system.

It is also essential to recognize that when groups of ions are separated by means of a selective precipitation, decantation after centrifuging will leave some of the solution along with the solid in the test tube. The Group I precipitate may be washed with a few drops of cold water to remove ions in Groups II and III. The Group II precipitate may be washed with 1 M NH_4Cl solution to remove Group III cations. The wash liquid should be thoroughly mixed with the solid to dilute the undesired cations. Centrifuge, decant, and discard the wash liquid.

Lead chloride is relatively soluble, so that although most of the Pb^{2+} will be removed in Group I, some PbS may precipitate in Group II. It will be dissolved along with the Bi^{3+} and Cu^{2+} in nitric acid, and will precipitate along with $Bi(OH)_3$. It will not, however, interfere with the test for Bi^{3+} that we have been using.

Since in this experiment you may be carrying out some reactions with which you are not familiar, use due caution and **remember to wear your safety glasses.**

Name _____ Section _____

ADVANCE STUDY ASSIGNMENT
AND LABORATORY REPORT: Laboratory Examination

Unknown no. _____

Possible ions _____   _____   _____   _____   _____

Flow Diagram

Ions found in unknown _____   _____   _____   _____   _____

Time turned in (to be entered by laboratory supervisor)   _____   **327**

APPENDIX I

Vapor Pressure of Water

Temperature °C	Pressure mm Hg	Temperature °C	Pressure mm Hg
0	4.6	26	25.2
1	4.9	27	26.7
2	5.3	28	28.3
3	5.7	29	30.0
4	6.1	30	31.8
5	6.5	31	33.7
6	7.0	32	35.7
7	7.5	33	37.7
8	8.0	34	39.9
9	8.6	35	42.2
10	9.2	40	55.3
11	9.8	45	71.9
12	10.5	50	92.5
13	11.2	55	118.0
14	12.0	60	149.4
15	12.8	65	187.5
16	13.6	70	233.7
17	14.5	75	289.1
18	15.5	80	355.1
19	16.5	85	433.6
20	17.5	90	525.8
21	18.7	95	633.9
22	19.8	97	682.1
23	21.1	99	733.2
24	22.4	100	760.0
25	23.8	101	787.6

To convert mm Hg to kPa, multiply the entry in the table by 0.1333.

$$1 \text{ mm Hg} = 0.1333 \text{ kPa}$$

APPENDIX II

Summary of Solubility Properties of Ions and Solids

	Cl$^-$	SO$_4$$^{2-}$	CO$_3$$^{2-}$, PO$_4$$^{3-}$	CrO$_4$$^{2-}$ yellow	OH$^-$ O^{2-}	H$_2$S, pH = 0.5	S^{2-}, pH = 9
Na$^+$, K$^+$, NH$_4$$^+$	S	S	S	S	S	S	S
Ba^{2+}	S	I	A	A	S$^-$	S	S
Ca^{2+}	S	S$^-$	A	S	S$^-$	S	S
Mg^{2+}	S	S	A	S	A	S	S
Fe^{3+} (yellow)	S	S	A	A	A	S	A
Cr^{3+} (blue-violet)	S	S	A	A	A	S	A
Al^{3+}	S	S	A, B	A, B	A, B	S	A, B
Ni^{2+} (green)	S	S	A, N	A, N	A, N	S	A$^+$, O$^+$
Co^{2+} (pink)	S	S	A	A	A	S	A$^+$, O$^+$
Zn^{2+}	S	S	A, B, N	A, B, N	A, B, N	S	A
Mn^{2+} (lt. pink)	S	S	A	A	A	S	A
Cu^{2+} (blue)	S	S	A, N	A, N	A, N	O	O
Cd^{2+}	S	S	A, N	A, N	A, N	A$^+$, O	A$^+$, O
Bi^{3+}	A	A	A	A	A	O	O
Hg^{2+}	S	S	A	A	A	O$^+$, C	O$^+$, C
Sn^{2+}, Sn^{4+}	A, B	A, B	A, B	A, B	A, B	A$^+$, C	A$^+$, C
Sb^{3+}	A, B	A, B	A, B	A, B	A, B	A$^+$, C	A$^+$, C
Ag$^+$	A$^+$, N	S$^-$, N	A, N	A, N	A, N	O	O
Pb^{2+}	HW, B, A$^+$	B	A, B	B	A, B	O	O
Hg$_2$$^{2+}$	O$^+$	S$^-$, A	A	A	A	O$^+$	O$^+$

Key: S, soluble in water.
    A, soluble in acid (6 M HCl or other nonprecipitating, nonoxidizing acid).
    B, soluble in 6 M NaOH.
    O, soluble in hot 6 M HNO$_3$.
    N, soluble in 6 M NH$_3$.
    I, insoluble in any common reagent.

S$^-$, slightly soluble in water.
A$^+$, soluble in 12 M HCl.
O$^+$, soluble in aqua regia.
C, soluble in 6 M NaOH containing excess S^{2-}.
HW, soluble in hot water.

Example: For Cu^{2+} and OH$^-$ the entry is A, N. This means that Cu(OH)$_2$(s), the product obtained when solutions containing Cu^{2+} and OH$^-$ are mixed, will dissolve to the extent of at least 0.1 mole per liter when treated with 6 M HCl or 6 M NH$_3$. Since 6 M HNO$_3$, 12 M HCl, and aqua regia are at least as strongly acidic as 6 M HCl, Cu(OH)$_2$(s) would also be soluble in those reagents.

APPENDIX III

Table of Atomic Masses
(Based on Carbon-12)

	Symbol	Atomic No.	Atomic Mass		Symbol	Atomic No.	Atomic Mass
Actinium	Ac	89	[227]*	Mercury	Hg	80	200.59
Aluminum	Al	13	26.9815	Molybdenum	Mo	42	95.94
Americium	Am	95	[243]	Neodymium	Nd	60	144.24
Antimony	Sb	51	121.75	Neon	Ne	10	20.183
Argon	Ar	18	39.948	Neptunium	Np	93	[237]
Arsenic	As	33	74.9216	Nickel	Ni	28	58.71
Astatine	At	85	[210]	Niobium	Nb	41	92.906
Barium	Ba	56	137.34	Nitrogen	N	7	14.0067
Berkelium	Bk	97	[247]	Nobelium	No	102	[253]
Beryllium	Be	4	9.0122	Osmium	Os	76	190.2
Bismuth	Bi	83	208.980	Oxygen	O	8	15.9994
Boron	B	5	10.811	Palladium	Pd	46	106.4
Bromine	Br	35	79.909	Phosphorus	P	15	30.9738
Cadmium	Cd	48	112.40	Platinum	Pt	78	195.09
Calcium	Ca	20	40.08	Plutonium	Pu	94	[242]
Californium	Cf	98	[249]	Polonium	Po	84	[210]
Carbon	C	6	12.01115	Potassium	K	19	39.102
Cerium	Ce	58	140.12	Praseodymium	Pr	59	140.907
Cesium	Cs	55	132.905	Promethium	Pm	61	[145]
Chlorine	Cl	17	35.453	Protactinium	Pa	91	[231]
Chromium	Cr	24	51.996	Radium	Ra	88	[226]
Cobalt	Co	27	58.9332	Radon	Rn	86	[222]
Copper	Cu	29	63.546	Rhenium	Re	75	186.2
Curium	Cm	96	[247]	Rhodium	Rh	45	102.905
Dysprosium	Dy	66	162.50	Rubidium	Rb	37	85.47
Einsteinium	Es	99	[254]	Ruthenium	Ru	44	101.07
Erbium	Er	68	167.26	Samarium	Sm	62	150.35
Europium	Eu	63	151.96	Scandium	Sc	21	44.956
Fermium	Fm	100	[253]	Selenium	Se	34	78.96
Fluorine	F	9	18.9984	Silicon	Si	14	28.086
Francium	Fr	87	[223]	Silver	Ag	47	107.870
Gadolinium	Gd	64	157.25	Sodium	Na	11	22.9898
Gallium	Ga	31	69.72	Strontium	Sr	38	87.62
Germanium	Ge	32	72.59	Sulfur	S	16	32.064
Gold	Au	79	196.967	Tantalum	Ta	73	180.948
Hafnium	Hf	72	178.49	Technetium	Tc	43	[99]
Helium	He	2	4.0026	Tellurium	Te	52	127.60
Holmium	Ho	67	164.930	Terbium	Tb	65	158.924
Hydrogen	H	1	1.00797	Thallium	Tl	81	204.37
Indium	In	49	114.82	Thorium	Th	90	232.038
Iodine	I	53	126.9044	Thulium	Tm	69	168.934
Iridium	Ir	77	192.2	Tin	Sn	50	118.69
Iron	Fe	26	55.847	Titanium	Ti	22	47.90
Krypton	Kr	36	83.80	Tungsten	W	74	183.85
Lanthanum	La	57	138.91	Uranium	U	92	238.03
Lawrencium	Lw	103	[257]	Vanadium	V	23	50.942
Lead	Pb	82	207.19	Xenon	Xe	54	131.30
Lithium	Li	3	6.939	Ytterbium	Yb	70	173.04
Lutetium	Lu	71	174.97	Yttrium	Y	39	88.905
Magnesium	Mg	12	24.312	Zinc	Zn	30	65.37
Manganese	Mn	25	54.9380	Zirconium	Zr	40	91.22
Mendelevium	Md	101	[256]				

*A value given in brackets denotes the mass number of the longest-lived or best-known isotope.

APPENDIX IV

Suggested Locker Equipment

2 beakers, 30 or 50 ml
2 beakers, 100 ml
2 beakers, 250 ml
2 beakers, 400 ml
1 beaker, 600 cc
2 Erlenmeyer flasks, 25 or 50 ml
2 Erlenmeyer flasks, 125 ml
2 Erlenmeyer flasks, 250 ml
1 graduated cylinder, 10 ml
1 graduated cylinder, 25 or 50 ml
1 funnel, long or short stem
1 thermometer
2 watch glasses, 3 or 4 in
1 crucible and cover, size #0
1 evaporating dish, small

2 medicine droppers
2 regular test tubes, 18 × 150 mm
8 small test tubes, 13 × 100 mm
4 micro test tubes, 10 × 75 mm
1 test tube brush
1 file
1 spatula
1 test tube holder, wire
1 test tube rack
1 tongs
1 sponge
1 towel
1 plastic wash bottle
1 casserole, small